POWDER METALLURGY PROCESSING
New Techniques and Analyses

CONTRIBUTORS

George E. Dieter
Richard W. Heckel
Robert M. Koerner
Howard A. Kuhn
Alan Lawley
John A. Tallmadge
Robert D. Thibodeau
Anthony R. Zecca

Powder Metallurgy Processing

NEW TECHNIQUES AND ANALYSES

Edited by

HOWARD A. KUHN

Department of Metallurgical and Materials Engineering
University of Pittsburgh
Pittsburgh, Pennsylvania

ALAN LAWLEY

Department of Materials Engineering
Drexel University
College of Engineering
Philadelphia, Pennsylvania

ACADEMIC PRESS   New York   San Francisco   London   1978
A Subsidiary of Harcourt Brace Jovanovich, Publishers

ACADEMIC PRESS, INC.
111 Fifth Avenue, New York, New York 10003

United Kingdom Edition published by
ACADEMIC PRESS, INC. (LONDON) LTD.
24/28 Oval Road, London NW1   7DX

Library of Congress Cataloging in Publication Data

Main entry under title:

Powder metallurgy processing.

Includes bibliographies.
1. Powder metallurgy.   I.   Kuhn, Howard A.
II.   Lawley, Alan.
TS245.P65   671.3'7   77-80789
ISBN 0-12-428450-7

CONTENTS

Chapter 1   POWDER PRODUCTION BY GAS AND WATER
             ATOMIZATION OF LIQUID METALS

John A. Tallmadge

Chapter 2   TRIAXIAL STRESS STATE COMPACTION OF
             POWDERS

Robert M. Koerner

Contents

LIST OF CONTRIBUTORS

Numbers in parentheses indicate the pages on which the authors' contributions begin.

GEORGE E. DIETER† (173), Processing Research Institute, Carnegie Mellon University, Pittsburgh, Pennsylvania

RICHARD W. HECKEL (51), Department of Metallurgical Engineering, Michigan Technological University, Houghton, Michigan

ROBERT M. KOERNER (33), Department of Civil Engineering, Drexel University, Philadelphia, Pennsylvania

HOWARD A. KUHN (99), Department of Metallurgical and Materials Engineering, University of Pittsburgh, Pittsburgh, Pennsylvania

ALAN LAWLEY (139), Department of Materials Engineering, Drexel University, Philadelphia, Pennsylvania

JOHN A. TALLMADGE (1), Department of Chemical Engineering, Drexel University, Philadelphia, Pennsylvania

ROBERT D. THIBODEAU (173), Special Projects Department, Atlantic Richfield Company, Los Angeles, California

ANTHONY R. ZECCA‡ (173), Research and Technology, Armco Steel Corporation, Middletown, Ohio

†Present address: College of Engineering, University of Maryland, College Park, Maryland.
‡Present address: Tax Department, Armco Steel Corporation, Middletown, Ohio.

PREFACE

In the last ten years, powder metallurgy and powder technology have made remarkable progress. The powder processing route is now recognized and respected as a competitive technology and an alternative to casting or conventional metal forming. Potential has also given way to proven capability in the production of new and unusual materials, parts, and components solely by powder fabrication.

Advances are reflected in (i) the conventional approach, in which the important steps are powder compaction and sintering and (ii) the newer sphere of high-performance powder metallurgy in which full density is achieved through hot isostatic pressing of loose powder or the working of a powder preform. In light of these latter developments, powder metallurgy can truly be considered to have "come of age." High performance powder metallurgy offers integrity of microstructure, compositional homogeneity, and mechanical property levels equal to (and frequently better than) those of the cast and wrought counterpart.

During this period of time, Drexel University received a major grant from the Department of Defense (under project "Themis") to conduct research and development studies in powder metallurgy. Recognizing the technical attraction of powder metallurgy for the production of metals and alloys in the form of semi-dense finished parts or fully-dense structural shapes, the major program goal was to hasten development of the necessary scientific and engineering base, i.e., to establish the powder fabrication route as a technologically and economically viable means of materials production. The outcome has been a series of studies in interrelated and intrarelated phases of powder metallurgy, each considered critical to the overall success of this form of material processing and fabrication. Areas selected were powder production by atomization, triaxial stress-state powder compaction, homogenization of metal powder alloys, the deformation and densification of porous sintered powder preforms, mechanical properties of high-density powder materials, and economic analysis of a representative industrial scale powder metallurgy operation. Each reflects the use of new analytic methods in metallurgy, mechanics, and economics in developing basic information on powder fabrication processes. The traditional processes of compaction and sintering were not included since they have received extensive treatment by other investigators; instead, innovative analyses and applications of these processes were developed, and are reflected in the homogenization and triaxial compaction areas.

The significance of the "Themis" program was subsequently reflected in interest and financial support from the National Science Foundation and industry. This enabled the group to further develop and apply these analytic techniques to a broad range of problems existing in powder processing. Companies involved were Alcoa, Ford Motor Company, Gleason Works and Hoeganaes Corporation, and International Nickel Company, Inc.

This book is the outcome of these studies. Each of the six major program areas is presented in the form of a self-contained chapter. A tutorial style has been adopted since the book is intended primarily for use in a first year graduate course in materials processing or as a self-contained primer in advanced powder metallurgy. The material covered should also be of assistance to practicing engineers in the powder metallurgy industry. Emphasis is placed on the fundamental phenomena involved, analytic techniques, and their application.

At the present time there is no text that goes beyond the introductory level in powder metallurgy processing. It is hoped that this new offering will, in part, fill this void and take the reader beyond a qualitative and descriptive treatment of the various powder processes. In summary, it is the intent of the text to demonstrate the application of advanced analytical techniques in metallurgy, mechanics, and economics to the fundamentals of powder fabrication processes.

ACKNOWLEDGMENTS

A close technical liaison was developed with the Frankford Arsenal as a result of the "Themis" program. We want to thank Harold Markus and Frank Zaleski for their constant encouragement, cooperation, and stimulation. Without their help, it would not have been possible to complete many of the phases of the overall powder metallurgy program.

A special debt of gratitude is due the several graduate students who worked with the faculty on this program. We were blessed with an unusually talented, innovative, and hard-working group. Former students were M. Balasub-ramaniam, C. L. Downey, D. L. Erich, B. L. Ferguson, R. J. Grandzol, M. M. Hagerty, A. J. Hickl, C. L. Jeanfils, R. D. Lanam, M. S. Masteller, W. M. McCabe, F. J. Quirus, P. Rao, S. K. Suh, R. A. Tanzilli, and A. R. Zecca.

Advice and constructive criticism were freely given by peers in academia and the powder metallurgy industries. In particular, we want to thank H. W. Antes, Eutectic Corporation (formerly with Hoeganaes Corporation); H. I. Aaronson, Michigan Technological University; K. E. Buchovecky, Alcoa; C. D. Durdaller, Hoeganaes Corporation; P. C. Eloff, Gleason Works; S. M. Kaufman, Ford Motor Company; M. J. Koczak, Drexel University; S. Mocarski, Ford Motor Company; D. W. Smith, Michigan Technological University; G. D. Smith, International Nickel Company, Inc.; C. W. Smith, Jr., Autoclave Engineering; and J. H. Tunderman, International Nickel Company, Inc.

Chapter 1

Powder Production by Gas and Water Atomization of Liquid Metals

John A. Tallmadge

DEPARTMENT OF CHEMICAL ENGINEERING
DREXEL UNIVERSITY
PHILADELPHIA, PENNSYLVANIA

I. INTRODUCTION

Processing of powder materials begins with consideration of the powder properties and characteristics, such as size, shape, composition, and struc-

1

ture. These powder features affect the response of subsequent compaction, sintering, and densification processes. Powder properties and characteristics are related to the method and conditions of production, and considerable effort is directed toward improved understanding of these processes to facilitate control of desired powder properties.

The various methods of powder production can be classified (Hirschhorn, 1969) as follows:

1. *atomization*, which may be used for any metal or alloy system that can be melted;

2. *chemical reduction*, as in the production of sponge iron from scale or iron oxides;

3. *mechanical crushing*, which is used for brittle materials such as beryllium and antimony; and

4. *electrolysis*, used for deposition of high purity powders.

Atomization and reduction processes are most widely used in high tonnage production, while mechanical crushing and electrolysis are used primarily for production of specialty materials in small quantities. The most flexible process is atomization since it provides the capability to produce alloy powders and affords greatest control over powder properties.

A. Atomization Overview

In general, atomization involves the formation of liquid metal droplets from a molten metal mass and subsequent or simultaneous solidification into powder particles. One type of atomization process uses a rotating disk or spindle to break up a liquid pool or stream into droplets, with cooling of the particles in an inert gas. By far the most widely used atomization process involves formation of solid metal particles from a molten metal stream by impingement with high velocity jets of another fluid (Gummeson, 1972). A typical atomization geometry is shown in Fig. 1. The major functions of the jets are

(i)   to break up the molten stream into small particles, and
(ii) to solidify the particles by quenching.

Quench atomization, as the process is called, thus involves the controlled flow of two fluids, which defines the analytical framework by which the process is studied.

Atomization processes for other types of materials include:

(a)   gas (air) aspiration, as used for making droplets of perfumes, aerosols, organics, and other solutions;

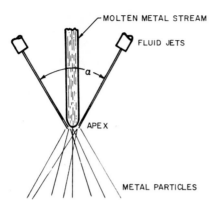

Fig. 1　Schematic of atomization geometry.

(b)　spinning disk processes, as used for evaporative spray drying of milk and coffee;

(c)　the one-fluid nozzle process, called pressure atomization, as used to form water droplets.

By comparison, quench atomization of metal powders involves the latent heat of fusion, whereas evaporative drying (b) involves the latent heat of evaporation, and (a) and (c) are nearly isothermal. Also by comparison, two-fluid quench atomization involves controlled flow of a second fluid (as does aspiration), whereas spinning disks and pressure nozzles involve controlled flow of only one fluid. The various processes involve different geometries and flow patterns, and thus lead to different fluid mechanics and heat transfer. Consequently, the mechanisms of two-fluid atomization differ from those of other types of atomization.

The high velocity fluid jets impinging on the molten stream may be water or gases. Water has higher viscosity, density, and quenching capacity than gases, but it also has an oxidizing tendency in atomization of reactive metals. Use of inert gases, in contrast, avoids oxidation. Water atomization processes also are less complex than gas atomization and require lower capital costs for atomization plant. Thus, water atomization is used predominantly for high-tonnage production of irregularly shaped iron-base alloy powders, while inert gas atomization is used for reactive and specialty alloy powders or where nearly spherical shape is required.

B.　Powder Parameters

Evaluation of the effects of process changes on powder characteristics requires techniques for measurement of particle parameters. The important characteristics of metal powders are included in the parameters:

1. *particle size* (such as mean size, size distribution, and yield);
2. *particle shape* (including related properties such as apparent density, flow rate, and surface area); and
3. *particle structure* (including density, chemistry, and grain size).

Quantitative description of particle size is readily accomplished by screen analysis, from which mean size, distribution, and yield can easily be calculated. Since most distributions of size versus cumulative weight fraction are log normal, they can be characterized by two parameters: (i) the midpoint (which gives d_m, the mean particle size) and (ii) the diameter ratio, or spread (which gives σ, the standard deviation). The latter is evaluated from the size for two cumulative mass fractions by

$$\sigma = d(0.841)/d(0.5) = d(0.5)/d(0.159) \tag{1}$$

where $d(x)$ is size d for a cumulative weight fraction x.

Simple, general quantitative measures of individual particle shape are not available. Because of the variety of shapes obtained, qualitative expressions such as *spherical* and *irregular* are often used, but neither of these is sufficiently descriptive. The standard tests which are available measure bulk powder properties, e.g., apparent density, flow time, and BET surface area.* These tests are of limited use because they are indirect measurements of shape and are influenced by parameters other than shape.

Characterization of structure is also complicated by the wide variety of structures that occur. Photomicrographs of mounted and polished particles are often used. Useful quantitative tests include particle density, composition, and hydrogen loss.

C. Process Parameters

Design and control variables in atomization processes may be manipulated to obtain desired powder characteristics. These parameters include:

1. *molten metal properties* (viscosity, surface tension, composition, superheat);
2. *molten metal flow geometry* (metal flow rate, stream length, flight path, manifold geometry);
3. *jet geometry* (apex angle, number of jets, jet location); and
4. *jet flow* (pressure, mass flow rate, viscosity, density).

---

*BET is a gas adsorption technique of surface area determination based on well-known theoretical principles put forth by Brunauer, Emmett, and Teller (1938).

Of particular importance in metal powder processing are the relationships between powder properties and the atomization parameters. The qualitative effects of many of these variables on particle size have been noted by Gummeson (1972), but little information has been published on the quantitative effect of most parameters.

This chapter focuses on the process of quench atomization and resulting powder characteristics. Experimental and theoretical studies, utilizing fluid mechanics and heat transfer principles, are described to elucidate the mechanisms of droplet formation and solidification and provide some indications of the relationships between process and powder parameters. Both gas and water are considered as the secondary fluid jets.

II. TYPICAL RESULTS OF JET ATOMIZATION

Before considering the mechanisms of atomization, it is useful to understand the relative importance of each of the many variables in atomization. This is accomplished by some preliminary experimental observations.

A. Experimental Apparatus and Procedure

The pilot atomization unit constructed for this study is shown in Fig. 2. The entire system was designed for operation at temperatures up to 1760°C (3200°F) to accommodate iron-base alloys. The quench chamber was fitted with a port to permit photography of the process. The tundish nozzles, which control metal flow rate, were 26 mm (1 in) long with inside diameters of 4.8, 6.4, and 8.0 mm (0.19, 0.25, and 0.31 in.). Jet nozzles for gas or water flow were threaded to one of three welded, four-sided manifolds. Each set of jet nozzles provides a different, but fixed, apex angle α (Fig. 1). A two-jet atomization geometry was used in most cases to facilitate photographic recording and mathematical modeling. Further details on pilot unit construction and results are given by Rao (1973) (gas atomization) and by Grandzol (1973) (water atomization).

Metal is melted in batches of 5 to 20 kg (11 to 44 lb) and then poured rapidly into the tundish where the temperature is measured. The metal flows through the nozzle at the bottom of the tundish, is atomized by high-pressure fluid from the manifold, quenched, and collected in a water pool at the bottom of the atomization chamber. After each run, the wet powders are dried with acetone and topped with a U.S. Standard Sieve (-40 mesh) to determine yield. The entire system is operated with good repeatability.

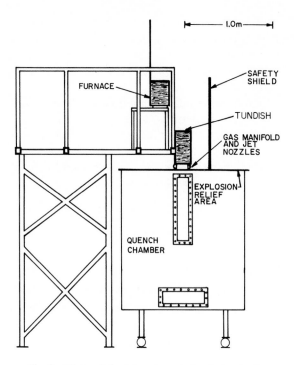

Fig. 2 Pilot unit for laboratory study of atomization.

B. Results

Atomization results are presented primarily as plots of particle size versus cumulative weight percent. The distributions are log normal, as indicated by the linear plots of size versus cumulative weight on log–log coordinates. This behavior was found for a wide range of conditions in both gas and water atomization, and has been reported by several previous investigators (e.g., Klar and Shafer, 1972). Consequently, the results can be expressed conveniently by the mean size d_m and diameter ratio (spread) σ, Eq. (1).

1. Gas Jets

Nitrogen is used as the atomizing fluid because it is relatively inert and low cost. Standard slot-shaped nozzles (Spraying Systems Model TTU) are used having dimensions of 1.16 by 7.8 mm (0.046 by 0.307 in.). The effects of gas pressure and jet distance (distance from the nozzle to the apex) on some properties of atomized cast iron powders are shown in Figs. 3 and 4 and Tables I and II.

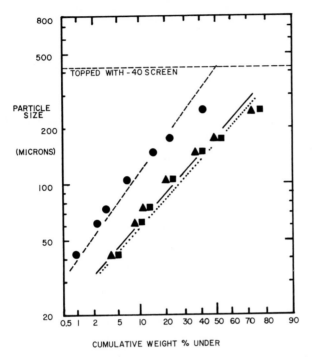

Fig. 3 Effect of jet pressure on powder size distribution in nitrogen atomization of cast iron. (Rao, 1973.)

	Pressure (kN/m²)	Run	Yield (wt%)	Mean size (μm)	σ
● — —	1383	C92	20.2	413	2.76
▲ —	2760	C93	79.1	177	2.33
■ · · ·	4140	C94	84.4	164	2.24

In gas atomization, the impinging fluid leaves the nozzle at high speed but slows down during its travel to the metal contact point (apex). The amount of this attenuation increases with increasing jet distance. For example, based on the nozzle exit velocity of 308 m/sec (1000 ft/sec) and jet distance of 73 mm (2.9 in.), the gas velocity at the apex was calculated from dispersion equations (Elrod, 1954) to be 62 m/sec (203 ft/sec). This attenuation decreases the jet fluid momentum available for breakup of the molten stream. Thus, increasing the gas pressure (increasing the jet nozzle exit velocity and, hence, the jet velocity at the apex) decreases particle size, as shown in Fig. 3. Similarly, decreasing jet distance (decreasing attentuation and, hence, increasing jet velocity at the apex) decreases particle size, as shown in Fig. 4.

In gas atomization, spherical particles are produced for most metals, but the cast iron of the present study produced cigar-like particles having

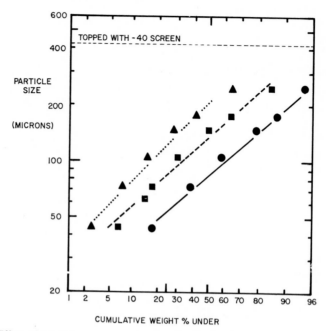

Fig. 4 Effect of jet distance on powder size distribution in nitrogen atomization of cast iron. (Rao, 1973.)

	Jet distance (mm)	Run	Yield (wt%)	Mean size (μm)	σ
▲ ···	99	C52	66.9	195	2.03
■ ----	73	C64	90.4	135	2.05
● ——	53	C51	92.0	82	1.91

large ratios of length-to-diameter (L/D). Measurements of L/D were obtained from macrographs of powder samples from the 100/140 mesh fraction. Fifty to eighty particles were measured in each sample and the results were averaged. Typical L/D values were between 2 and 3. Table II shows that L/D increases as jet distance increases.

Bulk properties of the powder properties are also given in Table II. Increasing deviation from sphericity (i.e., increasing L/D above 1.0) decreases apparent density and also increases flow time. This is as expected since increasing irregularity of the powder particles increases the interparticle frictional interference during motion of one particle over another.

2. Water Jets

Whereas gas atomization involves a diffuse, attenuating gas jet in a homogeneous gas phase, water atomization involves jets of small, high-velocity liquid droplets impacting the molten metal. These water droplets have two

TABLE I

Effect of Gas Pressure on Atomized Cast Iron Powder Properties[a]

Run number: Pressure, MPa (psig):	C92 1.38 (200)	C93 2.76 (400)
Size distribution (wt%)		
40/60	57.9	28.2
60/100	27.0	33.8
100/140	8.1	16.4
140/200	3.7	9.8
200/325	2.4	7.6
−325	0.9	4.2
Size characteristics (−40 mesh)		
Mean size (μm)	413	177
Diameter ratio, σ	2.76	2.33
Yield (wt%)	20	79
Properties of bulk sample (−40 mesh)		
Apparent density (g/cm³)	3.89	3.97
Flow time (sec)	32.8	27.4

[a]All runs at conditions of jet distance 73 mm (2.9 in), apex angle 60°, and batch size 7 kg (15 lb).

TABLE II

Effect of Gas Jet Distance on Atomized Cast Iron Powder Properties[a]

Run number: Jet distance, mm (in.):	C51 53 (2.1)	C64 73 (2.9)	C52 99 (3.9)
Size characteristics (−40 mesh)			
Mean size (μm)	82	135	195
Diameter ratio, σ	1.91	2.05	2.03
Yield (wt%)	92	90	67
Shape			
Mean L/D (100/140 fraction)	1.4	1.7	2.7
Properties of bulk sample (−40 mesh)			
Apparent density (g/cm³)	4.44	4.14	3.67
Flow time (sec)	16.0	20.6	31.6

[a]All runs at conditions of 4140 MPa (600 psig) pressure, 60° apex angle, and 7.7 kg (17.1 lb) batch size.

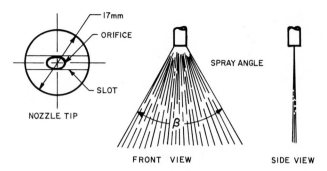

Fig. 5 Water nozzle geometry.

major advantages compared to gas atomization. First, they impact at a higher velocity because of higher jet nozzle exit velocity and negligible attenuation. Attenuation of the water jet droplets is negligible because, unlike the gas jets, they do not mix with the gas in the quench chamber and their higher momentum overcomes frictional drag. Second, water has a higher quench capacity due to both greater density and greater specific heat than gas.

Pressure atomization water nozzles were used to form the water droplet jets. An example is the flat fan spray nozzle, Fig. 5, which has a rectangular slot 4 by 17 mm (0.16 by 0.67 in.) with an inner elliptical orifice 4 by 6.5 mm (0.16 by 0.26 in.). With these jets, it was determined that the water jet velocity is constant over the jet distance and is related to nozzle pressure by

$$V = C_n P^{1/2} = 1.30 \, P^{1/2} \tag{2}$$

where V is jet velocity (meters per second) and P is nozzle pressure (kilonewtons per square meter). The actual coefficient C_n is, as expected, just below the theoretical value of $\sqrt{2}$ for frictionless flow through the nozzle. Visual observation of the nozzles indicated that water droplets formed within 5 mm (0.2 in.) of the nozzle tip. Thus, the water jet differs from the gas jet in terms of the fluid dynamics as well as heat transfer.

Water atomization trials were run using 4620 low alloy steel. The effect of water pressures on powder particle size is shown in Table III. Mean particle size decreases rapidly with increasing water pressure, i.e., increasing water jet velocity. Table III also shows that the spread, as indicated by diameter ratio, Eq. (1), is very nearly constant for each pressure level. Values of σ are in the range of 2.1 to 2.24.

Typical particles formed during water atomization are shown in Fig. 6. Increasing velocity (or nozzle pressure, Eq. (2)) increased the fraction of irregularly shaped particles.

TABLE III

Effect of Water Pressure on Atomized 4620 Steel Powder Powder Properties[a]

Run number:	87N	70N	61N
Pressure, MPa (psig):	1.72 (250)	3.45 (500)	13.8 (2000)
Size distribution (wt%)			
60/100	35.9	18.7	5.1
100/140	22.4	15.9	7.3
140/200	15.7	15.2	11.1
200/325	16.2	22.2	20.5
−325	9.8	27.0	56.0
Size characteristics (−60 mesh)			
Mean size (μm)	117	74	42
Diameter ratio, σ	2.10	2.24	2.15
Yield (wt%)	73	92	96

[a] All runs with two jets, jet angle 60°, jet distance 60 mm (2.3 in), fan spray angle 32° water flow rate 1.8 kg/sec (0.81 lb/sec), and batch size 8 kg (3.7 lb).

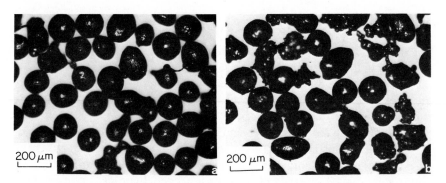

Fig. 6 Particle shapes produced by water atomization of 4620 low alloy steel (−100 + 140 mesh fraction). (a) Water jet velocity 50 m/sec (165 ft/sec). (b) Water jet velocity 149 m/sec 492 ft/sec).

III. MECHANISMS OF METAL DROPLET FORMATION

Several phenomena and principles are involved simultaneously in atomization processes: (i) complex fluid dynamics, (ii) heat transfer and energy conservation, and (iii) chemical kinetics and thermodynamics. The most important questions regarding mechanisms, however, are related to the fluid dynamics. To aid development of models of droplet formation, direct evidence is taken from photographs of the process rather than relying on indirect evidence, such as powder properties after quenching.

A. Gas Atomization

Stop-action still photographs of gas atomization were taken with a 35 mm camera and 55 mm lens at f 5.6 with an exposure time of 10^{-3} sec. An example, shown in Fig. 7, clearly indicates three metal zones:

   (i)   the upper, metal stream zone, in which standing waves are often observed in coherent streams and some stream separation is observed occasionally;
   (ii) the central, apex zone;
   (iii) the lower, shower zone in which individual particle trajectories are evident.

Such photographs are useful for studying oscillations in the upper, metal stream zone and velocities in the lower, shower zone.

Metal droplet velocities in the lower, shower zone appear to be a function of the distance z from the apex. Over the 0.1 to 0.4 m (4 to 16 in.) distance observed in the film, attenuation of the metal streak velocity u was correlated by

$$u = A \exp(-Bz) \tag{3}$$

For run C52, for example, A was found to be 18.5 and B to be 0.8, when u is expressed in units of meters per second and z in meters.

Figure 7 also shows the effect of gas pressure on the nature of flow below the apex. Increasing pressure appears to increase turbulence in the apex zone and finer particles result.

Although useful for the upper and lower zones, the still camera photographs are not satisfactory for stopping action in the central, or apex, zone where metal breakup and acceleration occurs. Because breakup is so important to fluid dynamics, study of the apex is needed. In order to stop action in the apex zone, a high-speed movie camera was used. With 1000 to 4000 frames/sec. exposure times varied from about 500 to 125 μsec/frame. A 7 mm Hycam camera was used with 121 m (400 ft) of 4X Negative 7224 Kodak film.

Using a 10 mm lens to provide a large field of view (300 by 900 mm) (1 by 3 ft), the high-speed movies showed the decomposition at the apex (Figs. 8a,b). With a 50 mm lens to provide a close-up field of view (50 by 150 mm) (2 by 6 in), the high-speed movies showed single and multiple ligaments being formed and breaking into individual droplets (Figs. 8c,d.).

Photographic studies of gas atomization have also been reported by See *et al.* (1973). Using a high-speed camera and floodlight or stroboscopic illumination, the atomization of liquid lead was studied at a metal temperature of 413°C (858°F). Results illustrated the hollow metal cone formed just

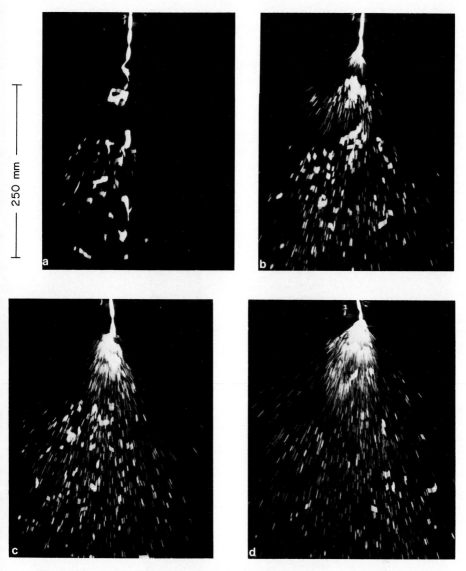

Fig. 7 Photographs of gas atomization of cast iron, showing effect of jet pressure. (a) 0.345 MPa (50 psig). (b) 0.7 MPa (100 psig). (c) 1.045 MPa (150 psig). (d) 1.4 MPa (200 psig). Other conditions: jet distance 76 mm (3 in), jet angle 45°, metal temperature 1700°K (3911°F), run C24.

above the geometrical impingement point. The cone underwent oscillations and instabilities leading to cyclic behavior of the metal stream. A similar result was observed by Rao (1973) as shown in Fig. 7.

See *et al.* (1973) related breakup of the cone to previous studies of breakup of flat sheet by Fraser *et al.* (1962) and Dombrowski and Johns (1963). The latter mechanism involves a five-step sequence:

(1)   growth of waves in the sheet;
(2)   breakup of the sheet into ligaments;
(3)   growth of thickness variations along the ligament length;
(4)   breakup of the ligament into droplets; and
(5)   spheroidization of the droplets.

This model will be discussed more fully in Section IV.

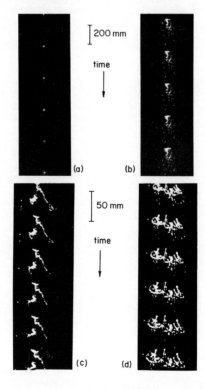

Fig. 8   High-speed movies of gas atomization (2000 frames/sec). (a) Jet distance 73 mm (2.9 in). (b) Jet distance 99 mm (3.9 in). (c) Single ligament breakup. (d) Multiparticle breakup (jet pressure 2.08 MPa (300 psig), jet angle 26°, metal temperature 1700°K (3911°F)).

B. Water Atomization

During water atomization, three zones are observed (metal stream, apex, and shower zones) similar to gas atomization. While in gas atomization solidification most likely occurs in the shower zone, in water atomization solidification probably occurs in the apex zone because of the higher quench capacity of water. This leads to greater occurrence of irregularly shaped particles during water atomization. Thus, photography of the apex zone is of most interest in water atomization.

Grandzol (1973) found that photographs of water atomization were difficult to obtain with ordinary film because of fog and absorbed water vapor on the quench chamber portal (Fig. 2). Infrared film, however, gave clear outlines of the oscillatory shape of the metal stream above the apex. The apex zone was unclear but some discrete streaks were observable in the shower zone. Use of a rotating shutter, which reduced exposure time to 10^{-4} sec, did not improve clarity.

In one run, with a water jet velocity of 50 m/sec(165 ft/sec), metal particle velocities measured from the infrared photographs were 15.3 m/sec (50.5 ft/sec). Particle velocities were approximately independent of particle diameter in the observable range of sizes.

Based on the visual, photographic, and powder property observations in this study, it is clear that the two basic steps involved in atomization are breakup and quenching. As summarized in Tables IV A and IV B, breakup leads to droplet formation and affects particle size, while quenching leads to solid particle formation and affects shape.

This section primarily considered the fluid dynamics of atomization through photographic evidence. Heat transfer phenomena are also important and will be treated in a more theoretical manner in the next section. The third type of phenomena referred to in the opening paragraph of this section, chemical reactions and thermodynamics, deserve attention because of the importance of oxygen content and other properties; however, this study has not addressed chemical reactions and little or nothing has been published on this subject by previous investigators.

IV. THEORETICAL CONSIDERATIONS

As illustrated in Tables IV A and IV B, breakup of the molten metal stream into droplets involves fluid dynamics, and solidification of the droplets into powder particles involves heat transfer. The former controls particle size

TABLE IVA

Two Basic Steps Observed

Molten metal stream	$\longrightarrow$	1. Particle size formation (breakup)	$\longrightarrow$	Molten droplets	$\longrightarrow$	2. Solid shape formation (quenching)	$\longrightarrow$	Solid metal powder
Results:		(a) Particle size				(a) Same size		
		(b) Initial shape				(b) Final shape		
Mechanisms:		(a) Breakup (initial)				(a) Cooling		
		(b) Breakup (secondary)				(b) Surface smoothing		
		(c) Coalescence				(c) Oscillations		

TABLE IVB

Special Mechanisms Observed

Type of atomization	Basic step	
	1. Particle size formation	2. Solid shape formation
Water atomization mechanisms	(a) Initial breakup due to impact of water droplets	(a) Rough surfaces retain for shorter solidification times
	(b) Initial shape influenced by water evaporation	(b) Smooth surfaces occur with longer solidification times
	(c) Little coalescence	
Gas atomization mechanisms	(a) Breakup via ligament decomposition	(a) Oscillations may occur
	(b) Little Coalescence	(b) Droplet length to diameter (L/D) may change

and the latter controls particle shape. In this section, fluid dynamics and heat transfer analyses are given for both gas and water atomization.

A. Fluid Dynamics of Gas Atomization

As described in Section III.A, the liquid stream breakup process involves formation of ligaments and subsequent separation of the ligaments into droplets. Breakup of thin cylinders was first described by Rayleigh (1898) for inviscid flow, who found that

$$d = 1.88d_{\text{L}} \qquad (4)$$

where d is the droplet diameter and d_L is the ligament diameter. See *et al.* (1973) presented a correction to Eq. (4)

$$d = 1.88 d_L (1 + 3\varphi)^{1/6} \tag{5}$$

where ϕ is a function of $d_L^{-1/2}$, liquid viscosity, density, and surface tension. Because d_L appears as $d_L^{1/12}$ in the correction factor of Eq. (5), however, the relationship between d and d_L is adequately given by Eq. (4).

See *et al.* (1973) also presented an equation for ligament diameter formed from a sheet of liquid

$$d_L = 0.961 (h^2 x^2 \phi_1/u^4)^{1/6} \left[1 + 2.6 (h x u^7 \phi_2)^{1/2}\right]^{1/5} \tag{6}$$

where h is sheet thickness, u the relative velocity between the gas and liquid, x the distance along the sheet, and ϕ_1 and ϕ_2 are functions of gas and liquid metal properties. Experimental data presented in Figs. 3 and 4, as well as by See *et al.* (1973) and Small and Bruce (1968), indicate that increasing velocity decreases particle size. Thus, the second term involving velocity u in Eq. (6) must be small compared with unity, and, neglecting this term, Eq. (6) reduces to

$$d_L = \frac{0.961 \left[h^2 x^2 \gamma^2/\rho_g \rho_l\right]^{1/6}}{u^{2/3}} \tag{7}$$

where γ is the surface tension between the liquid and gas, ρ_l the liquid metal density, and ρ_g the gas density.

Combining Eqs. (4) and (7), the predicted particle size is

$$d = \frac{1.81 \left[h^2 x^2 \gamma^2/\rho_g \rho_l\right]^{1/3}}{u^{2/3}} \tag{8}$$

In experiments on nitrogen atomization of lead, See *et al.* (1973) measured a particle size from 305 μm to 180 μm with a velocity increase from 10.1 to 15.6 m/sec (33.2 to 50.6 ft/sec). This data conforms with a velocity factor of $u^{-1.2}$ rather than $u^{-0.67}$, as predicted in Eq. (8). Nevertheless, Eq. (8) seems to be a useful qualitative contribution worthy of further study.

Another shearing model begins with the equation for ligament diameter due to Dombrowski and Hooper (1962)

$$d_L = 3(\lambda h)^{1/2} \tag{9}$$

where h is the sheet thickness and λ is the wavelength of sheet breakup. This is combined with the equation for wavelength due to Squire (1953)

$$\lambda = 3\pi\gamma/\rho_g u \tag{10}$$

Combining these relationships with the Rayleigh relationship, Eq. (4), gives

$$d = 17.3(\gamma h/\rho_g u)^{1/2} \tag{11}$$

which predicts the effect of velocity as $u^{-0.5}$. Further work is required to establish the velocity exponent and the quantitative influence of physical properties on mean particle size.

B. Heat Transfer in Gas Atomization

In Section II.B.1, the shape parameter L/D was defined to illustrate the effect of gas pressure on shape. Based on these experimental results, Rao (1973) has developed the only known model for predicting particle shape. Two submodels were developed for solidification time (as a function of particle velocity, size, superheat, and fluid properties) and for shape change with time (as a function of initial shape, drag, and fluid properties).

The solidification model was based on heat transfer during convection cooling of a sphere in flight; radiation and other flow effects were shown previously by Rao (1970) to be negligible. Combining the convection rate with the transient energy balance for a particle leads to the expression for solidification time

$$t_s = (\rho C_p D_0/H) \ln\left[(Tm - Tg)/(Ts - Tg)\right] \tag{12}$$

where ρ is the metal density, C_p the specific heat of the liquid metal, D_0 the particle diameter, H the convection coefficient, T_m, T_g, and T_s the temperatures of liquid metal, atomizing gas, and solidification, respectively.

The shape change model was based on control-volume force balance for an elongated, ligament-shaped particle. Four forces were considered: (i) surface tension, (ii) viscous damping, (iii) inertia, and (iv) surface drag. Drag was shown to have a small effect and, for simplicity, was neglected.

The resulting differential equation (Rao, 1973) predicted oscillations with an exponential decay function and characteristic time t_c. Characteristic time is a function of particle properties, including size, and liquid metal viscosity. The result is prediction of the shape parameter as a function of time

$$L/D = 1 + (L_0/D_0 - 1)\ \exp\left[-t/t_c(\gamma/m)\right]^{1/2} \tag{13}$$

where m is the particle mass and L_0/D_0 is the initial length-to-diameter ratio of the ligament. Setting time t in Eq. (13) equal to the solidification time t_s, Eq. (12), the solidified particle shape parameter L/D can be calculated.

Using the experimental parameters for nitrogen atomization of cast iron and assumed initial L_0/D_0 values of 2.2, 2.7, and 3.2, predicted particle

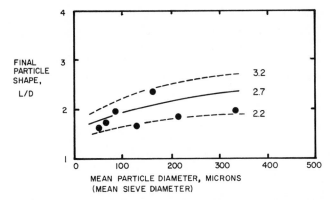

Fig. 9 Variation of particle shape with particle size in gas atomization of cast iron; comparison of theoretical prediction, Eqs. (12) and (13) (——, ———) with experimental data (●). Jet distance 73 mm (2.9 in), jet pressure 4.14 MPa (600 psig), jet velocity 62 m/sec (205 ft/sec); initial shapes (L_0/D_0) for predicted curves are 3.2, 2.1, 2.2.) (Rao, 1973.)

shapes were calculated as a function of particle size. For one particular set of conditions, these relationships are shown in Fig. 9. This is the first known theoretical prediction of particle shape. Additional breakup models predicting initial shape L_0/D_0 would allow calculation of shape from primary atomization variables. Considerable further study is required, however, before such models can be developed.

C. Fluid Dynamics of Water Atomization

Breakup of a liquid metal stream by impact of high-velocity liquid droplets is more difficult to describe than gas atomization because a gas phase exists in addition to the liquid metal and atomizing liquid stream. There is limited evidence from high-speed photographs that a liquid metal cone forms periodically, from which sheets of liquid are torn away and decomposed into ligaments and then droplets. If this result is general, then breakup during water atomization is basically an accelerated process related to gas atomization.

There are two potential models of metal breakup in water atomization. The first would involve a shearing mechanism of a metal stream or cone, based on the Rayleigh model, as discussed by Grandzol (1973). Such a model has not been developed, but it would be expected to give a predicted relationship for particle size similar to that for gas atomization, Eq. (8).

A second model due to Grandzol and Tallmadge (1973) is based on the observation that water droplets make up the water stream impacting the metal stream. Principles of momentum and energy conservation are used along with experimental evidence obtained photographically. Consider a water droplet impacting the metal stream surface and causing a number n of metal droplets to be formed. As a first approximation, assume that the water droplet transfers all of its momentum to the newly formed metal droplets, i.e., $M_m = M_w$, where M is momentum and subscripts m and w refer to metal and water droplets, respectively. Expressing momentum in terms of particle diameter d, velocity V, and density ρ leads to

$$d_m = d_w (V_w \rho_w / n V_m \rho_m)^{1/3} \qquad (14)$$

Equation (14) predicts that particle size is a function of water droplet size and metal particle velocity. Determination of water droplet size involves equations for one-fluid pressure atomization, for which general empirical correlations of the form $d_w = B/V_w$ are available. Metal particle velocity was determined photographically for a specific run with a known water velocity. A kinetic energy balance suggests that the ratio of water droplet velocity to metal particle velocity is constant for all velocities. This ratio was determined from photographic data to be $V_w/V_m = 3.37$. Substituting these relationships into Eq. (14) gives

$$d_m = (Bn^{-1/3}/V_w)(3.37 \, \rho_w/\rho_m)^{1/3} \qquad (15)$$

Including the densities of water and liquid 4620 steel and the value of B determined by Merrington and Richardson (1947), Eq. (15) becomes

$$d_m = 14,900 \, n^{-1/3}/V_w \qquad (16)$$

Substituting Eq. (2) relating water jet velocity to nozzle pressure, the expression for particle size becomes

$$d_m = 11,461 n^{-1/3} P^{-1/2} \qquad (17)$$

In this section, dynamic breakup models for prediction of particle size have been developed for gas and water atomization. A fluid dynamics–heat transfer model of shape formation in gas atomization has also been presented. Heat transfer models for quenching in water atomization are much more complex because much of the cooling occurs simultaneously with breakup and particle formation in the apex zone, rather than sequentially, as in gas atomization.

V. EXPERIMENTAL RESULTS

Sections II and III provided an experimental overview of atomization, illustrating the major parameters and suggesting some plausible mechanisms. Section IV gave further insight into the mechanisms through fluid dynamics and heat transfer analyses. In this section, further experimental results are given for evaluation of the analytical models of particle size and shape.

A. Water Atomization

1. Size

A total of 22 pilot runs were made on water atomization of 4620 steel. A two-jet configuration was used with stainless steel, flat fan jets having spray angles (β in Fig. 5) of 25° and 32°.

The molten metal temperatures were measured by immersing a platinum–rhodium thermocouple into the tundish. Temperatures for 4620 steel were held approximately constant near 1710°C (3110°F) usually within ±20°C (±36°F) and the variation in metal flow rates, for a given nozzle diameter, was held to ±10%. Metal flow rates were varied by changing the diameters of the tundish nozzles.

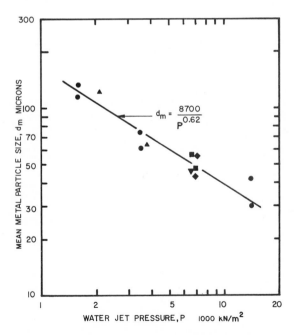

$$d_m = \frac{8700}{P^{0.62}}$$

Fig. 10 Correlation between mean particle size and jet pressure in water atomization of 4620 steel. Water jet flow rate (kg/sec): ▲, 0.9; ■, 1.3; ●, 1.8; ◆, 2.5; ▼, 3.8. (Grandzol, 1973.)

Conditions measured directly were water flow rate, water pressure, metal temperature, metal batch size, and time of metal flow. From these experimental quantities, metal flow rate, water jet velocity, water jet momentum, and water jet kinetic energy were calculated.

It was found that the size distribution did not vary appreciably with water velocity, water flow rate, or metal flow rate over the ranges studied. Furthermore, it appeared that the spread for all runs was approximately constant at a diameter ratio σ of about 2.25 $\pm$ 10%, which is within experimental error. It was concluded, therefore, that the spread of size distributions may be considered to be constant for the conditions studied. This is an important simplification, which allowed description of size data using a single parameter, the mean size.

The effect of pressure on size was studied first, because pressure is a primary control parameter in commercial practice. As shown in Fig. 10, particle size decreases with increasing pressure, as expected (see Section II.B.2). The slope of the line through the data, i.e., the exponent on pressure, has a value of -0.62, while Eq. (17) predicts a slope of -0.5.

The problem with this test, however, is that flow rate and velocity both change with pressure. From a fundamental standpoint, it is desirable to study the effect of flow rate and velocity independently. Thus, various

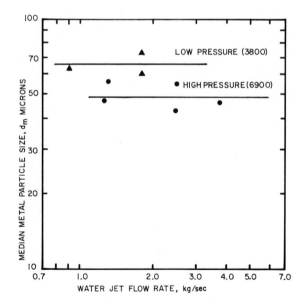

Fig. 11   Effect of jet flow rate on particle size in water atomization of 4620 steel. Pressure (kN/m^2): ▲, 3800; ●, 6900. Water velocity (m/sec): ▲, 79; ●, 107. (Grandzol, 1973.)

nozzles were used to study change of velocity at constant flow rate and vise versa.

The influence of water jet flow rate on metal particle size was plotted at two pressures (i.e., two velocities). As shown in Fig. 11, water jet flow rate had no appreciable effect on metal particle size. This small or negligible effect of flow rate is also indicated in Fig. 10, where the data scatter is small despite the fourfold variation in water flow rate.

The influence of water jet velocity at different flow rate conditions is shown in Fig. 12. The correlation is good and the arrangement of data points is almost identical to that of the pressure plot, Fig. 10. This is reasonable since water droplet velocity and water pressure are directly related, as shown in Eq. (2), and since water jet flow rate did not have any effect on particle size. The slope of the line in Fig. 12 is about -1.0. The equation of the line drawn through the points is

$$d_\mathrm{m} = 5500/V_\mathrm{w} \tag{18}$$

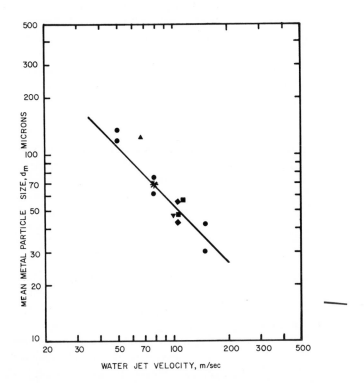

Fig. 12 Effect of jet velocity on particle size in water atomization of 4620 steel. Water flow rate (kg/sec): ●, 0.9; ▲, 1.3; ■, 1.8; ◆, 2.5; ▼, 3.8. (Grandzol, 1973.)

This compares favorably with model Eq. (16) of Section IV.C. The effect of jet distance was studied at constant velocity of 79 m/sec (260 ft/sec) and constant water flow rate 0.91 kg/sec (1.98 lb/sec). Although jet distance was reduced from the normal length of 120 to 60 mm (4.7 to 2.35 in) the mean sizes were nearly identical (64 and 67 μm, respectively) and the powders were virtually indistinguishable. Based on the constancy of jet velocity with jet length for water jets, these data also confirm the importance of jet velocity (or pressure) as the fundamental variable.

The fundamental dependence of particle size on velocity is an important concept because the pressure parameter is reported in most previous work. Pressure does not appear to be a fundamental parameter, partly because it describes the water condition upstream of the nozzle rather than at the metal apex.

Pilot runs at apex angles (see Fig. 1) of 30°, 45°, and 60°, holding other conditions constant, resulted in a decreasing median size of 91, 80, and 64 μm, respectively. This trend is consistent with that reported by Gummeson (1972). The influence of the apex angle α between the two jets was studied vectorially by considering the influence of $\cos(\alpha/2)$ and $\sin(\alpha/2)$ graphically (Grandzol and Tallmadge, 1975). It was found that the normal velocity component of the sine term drew the data together much better than the tangential velocity component of the cosine term. The size correlation was, therefore, modified to

$$d_{\mathrm{m}} = \frac{2750}{V_{\mathrm{w}}\sin(\alpha/2)} \tag{19}$$

Other parameters were varied but, within the range studied, no appreciable influence on mean size was observed. These parameters are metal flow rate, ratio of water to metal flow rates, and fan spray angle.

Three or four runs were made with each of several other metals using an apex angle of 60°. In each case the mean size data are satisfactorily described by

$$d_{\mathrm{m}} = A/V_{\mathrm{w}} \tag{20}$$

Values of the parameter A for each metal are given in Table V together with some of the metal properties. The mean size are very similar ($\pm 20\%$) for metals with a wide range of melting points. Shapes were also studied photographically and found to vary apparently with surface tension.

An exploratory study of the influence of fluid properties on size, based on the metals in Table V, indicates that mean size may increase with viscosity but is insensitive to surface tension. This is an area requiring further study.

Kishidaka (1972) has presented a generalized, empirical, size correlation

TABLE V

Mean Particle Sizes for Water Atomization

Metal	4620 Steel	304 Stainless steel	Cast iron	Copper	Bismuth alloy Cerrobend
Composition rated (wt%)	1.8 Ni, 0.55 Mn, 0.25 Mo, 0.25 Si, 0.20 C, bal. Fe	19 Cr, 9 Ni, 0.08 C, bal. Fe	3.5 C, 2.0 Si, bal. Fe	99.9 Cu	50 Bi, 27 Pb, 13 Sn, 10 Cd,
Atomization conditions					
Tundish temp. (°C)	1710	1650	1380	1240	175
Melting pt. (°C)	1516	1413	1177	1082	70
Superheat (°C)	190	240	200	160	110
Properties					
Surface tension (N/m)	1.8	1.8	1.0	1.4	0.5
Viscosity (Ns/m²) × 10⁻³	2.5	2.5	1.4	3.9	2.5
Mean size A parameter Eq. (20)	5500	5000	3700	3800	4600

for two-jet water atomization of iron

$$d_m/D = a\text{Re}^b\text{We}^c(W_w/W_l)^d \tag{21}$$

where D is tundish nozzle (or metal stream) diameter, Reynolds number $\text{Re} \equiv D\rho_l V_w/\mu_l$, Weber number $\text{We} \equiv D\rho_l V_w^2/\gamma_l$, W is mass flow rate, V_w is water droplet velocity, μ is viscosity, ρ is density, subscript l refers to the liquid metal, subscript w to water, and a, b, c, and d are constants.

Kishidaka (1972) reported values of $b = -0.57, c = -0.22$, and $d = -0.43$. Because of the small exponent on W_w, the influence of water flow rate is almost negligible. Grandzol (1973) has shown that Eq. (21) can be simplified by eliminating stream diameter and mass flow rates to obtain

$$d_m = (a/V_w^{1.01})(\gamma^{0.22}\mu_l^{0.57}/\rho_l^{0.79}) \tag{22}$$

For atomization of pure iron, this reduces to

$$d_m = 9,000/V_w \tag{23}$$

Equations (22) and (23) show agreement in terms of the effect of water velocity with the empirical equation (18) and model equation (16).

2. Shape

Very few systematic and quantitative studies of the effect of water jet parameters on powder shape have been published. One of the few is that of Grandzol (1973), who did exploratory work on factors involving shape. The first task was to define shape quantitatively for powders such as shown in Fig. 6. It was noted that two different types of shape (smooth and spherical, rough and irregular) appeared in these and other samples, so that these shapes did not lend themselves to shape characterization parameters used previously, such as sphericity or length-to-diameter ratio. Grandzol (1973) defined shape by the fraction (F_i) of particles which were rough and irregular, as opposed to those which were smooth and spherical. The rough surface is indicative of a rapid quench, with some irregularity due perhaps to coalescence of particles.

A plot of rough fraction F_i versus size for 4620 steel powders, Fig. 13, indicated a significant decrease in irregular shape with larger sizes. Specifically, the irregular fraction decreased from about 80% at 40 μm to about 30% at 120 μm.

When rough fraction F_i is plotted versus jet pressure (or jet velocity) the scatter was greater than with Fig. 13, but increasing pressure did result in a higher rough fraction, as expected. Changes in water jet flow rate at constant pressure, on the other hand, did not appear to change the rough fraction.

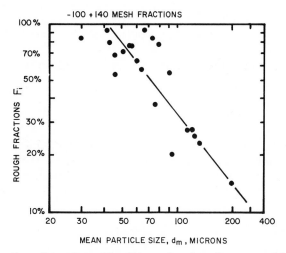

Fig. 13 Fraction of rough particles F_i as a function of mean particle size in water atomization for 4620 steel. (Grandzol, 1973.)

Gummeson (1972) reported that spherical particles were favored by larger apex angles. A similar study by Grandzol (1973) was inconclusive, perhaps due to scatter. For example, for angles of 30°, 45°, and 60°, rough fractions were 54, 77, and 57%, respectively. However, if the effects of velocity and apex angle are considered together, then the rough fraction results are consistent with Gummeson's report.

Summarizing the results of this section, data obtained using 4620 steel and four other metals showed that the mean particle size depends primarily on water velocity and jet angle but is not influenced noticeably by changes in water flow rate or other variables. The observed inverse proportionality between particle size and jet velocity is consistent with the predictions of a theoretical model of water atomization.

B. Gas Atomization

In Section IV.B, a unique fluid dynamics–heat transfer model was given for prediction of particle shape in gas atomization. Length-to-diameter ratio L/D was used to characterize shape. These results were compared with experimental results from 32 pilot runs on cast iron. This alloy was chosen because it resulted in powder particles with large L/D values.

In one series of ruler measurements from photographs of 75 particles for a 100/140 mesh fraction, L/D ratios varied from 1.06 to 5.44 for individual particles. The arithmetic mean and standard deviation (σ) from the mean were

$$L/D = 2.60 \pm 1.08$$

In two repeat tests of the same powder, mean values were found to be

$$L/D = 2.61 \pm 1.10, \qquad L/D = 2.69 \pm 0.98$$

Since the average ruler measurement error is about 8% in L/D, the L/D technique was considered repeatable and reliable.

For correlation purposes, L/D values were compared with bulk properties, apparent density, and flow time determined using ASTM tests. Powders were fractionated into 60/80, 100/140, and 140/200 mesh fractions. Plots of flow time versus L/D showed a linear and increasing relationship, for each size, given by

$$t = b + a(L/D)$$

where t is flow time in seconds, a the slope, and b the extrapolated intercept. The slope a was the same (6.5) for all three sizes, and the intercept b varied from 16 (for the 60/80 mesh) to 10.5 and 6.5 for the 100/140 and 140/100 fractions.

Similarly, plots of apparent density and tap density versus L/D showed linear, decreasing relationships. Expressing density in units of kilograms per cubic meter, the slope for apparent density for all three size fractions was -470, and the intercepts were 4600, 4720, and 4910 for the size fractions 60/80, 100/140, and 140/200, respectively. The slope for tap density was -220 for all three sizes, and the intercepts were 4810, 4920, and 4970, respectively.

The effect of size on shape of gas atomized powders was also studied from photographs of close size fractions ranging from 230 mesh (53 μm) to 80 mesh (177 μm). As shown in Fig. 14, smaller particles are more spherical while larger particles have larger L/D ratios. Very large particles in the 40/80 mesh range (up to 420 μm) were also flatter, probably because of impact with chamber walls or the water pool at the bottom of the quench chamber. Variation of apparent density and flow time with particle size, Figs. 15 and 16, also show the expected effects of size.

Having established L/D as a proper shape characterization and shown its correlation with bulk properties, a systematic study of atomization jet parameters was conducted. Parameters used to characterize shape were L/D for the 100/140 size fraction and apparent density and flow time for the entire -40 mesh sample.

The effect of nozzle pressure on shape is substantial, with average L/D decreasing from 2.7 at 1.4 MPa (200 psig) to 2.4 at 2.8 MPa (400 psig) to 1.7 at 4.1 MPa (600 psig). Bulk properties showed a similar consistent trend

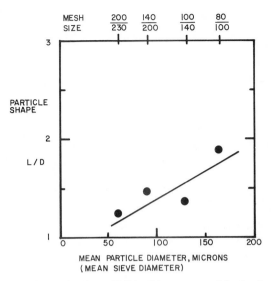

Fig. 14 Variation of particle shape (*L/D*) with mean particle size in gas atomization. Run C51. (Rao, 1973.)

toward sphericity (lower flow time, higher apparent density) at higher pressure. This large, inverse effect of pressure on irregular shape was expected and gives a quantitative description of the effects of pressure.

It was shown in Section II that jet distance has a substantial effect on powder shape. Referring to Table II, the shape parameter *L/D* increases from 1.4 at the jet distance of 53 mm (2.1 in) to 2.7 at 99 mm (3.9 in). This result is unexpected, and no explanation is currently available.

Gas flow rate was varied by using a large nozzle, with an orifice 2.3 by 7.8 mm (0.09 by 0.31 in). This doubled the gas flow rate to 0.36 kg/sec (0.78 lb/sec) at the same pressure. As shown by runs C94 and C95 in Table VI, the influence of gas flow rate on shape was found to be small, but noticeable for the twofold range studied. Increasing gas flow rate increases sphericity, which is consistent with the pressure effect noted previously.

Variation of particle shape with particle size was shown in Fig. 9 for comparison with calculated values from Eqs. (12) and (13) in Section IV.B. These data suggest that the initial shape parameter L_0/D_0 is in the range of 2.2 to 3.2. The model properly predicts the influence of particle size on shape. Comparisons between the model and experimental data for other conditions, however, indicate some discrepancies. In its current form, the model is limited, but worthy of further study.

Gas atomization of 4620 low alloy steel, 1020 carbon steel, 304 stainless steel, 718 Inconel, and copper led to nearly spherical particle shapes. For example, mean L/D values of 1.2 and 1.4 were measured in two stainless steel runs and values of 1.2 and 1.3 in copper runs. Reasons for the lack of ligaments in these powders are not clear, but it is believed that fluid properties of the molten metal are influential.

This chapter has considered several jet variables and other parameters which influence powder properties. Attention was directed to the most important parameters involved in atomization mechanisms, but an extremely important aspect not discussed here is the influence of major changes in jet geometry (design and layout). The choice of jet geometry is very important for commercial production and is the subject of much of the patent literature in this field. Two excellent references on jet design are Gummeson (1972) for water jets and Klar and Shafer (1972) for gas jets.

As noted previously, little attention was given to problems of structure, including oxygen content, mass transfer, diffusion, chemical kinetics, and grain structure. This is not because of lack of importance but due to an apparent lack of fundamental work in this area.

VI. SUMMARY AND CONCLUSIONS

In water atomization of ferrous and nonferrous metals, it has been found that mean particle size depends primarily on water velocity and jet angle but is not significantly affected by water flow rate. This response was predicted theoretically using a droplet impact model. The influence of atomization parameters on particle shape and other powder properties was also studied.

From experimental studies of the gas atomization of cast iron, it is concluded that particle shape is controlled primarily by particle size and jet distance but is not influenced to any extent by superheat or other processing variables. A shape model based on quench-time and shape-time models

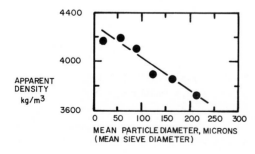

Fig. 15 Decrease in apparent density with increasing size of gas atomized particles of cast iron. Run C51. (Rao, 1973.)

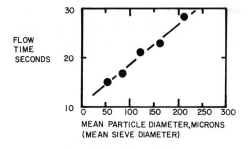

FLOW
TIME
SECONDS

MEAN PARTICLE DIAMETER,MICRONS
(MEAN SIEVE DIAMETER)

Fig. 16 Flow time increase with increasing size of gas atomized particles of cast iron. Run C51. (Rao, 1973.)

predicts this behavior. High-speed and stop-action photography were used to study breakup mechanisms.

It is believed that these studies indentify, for two-jet atomization, the fundamental variables of primary importance and those of secondary importance in water and gas atomization. From observation of particle formation

TABLE VI

Conditions and Powder Properties for Gas Atomization of Cast Iron[a]

	Shape of one size fraction Mean L/D of 100/140 fraction	Properties of bulk sample (−40)	
		Apparent density (g/cm³)	Flow time (sec)
Gas flow rate			
0.19 kg/sec (0.42 lb/sec) (run C94)	1.7	4.23	23.6
0.36 kg/sec (0.79 lb/sec) (run C95)	1.5	4.53	17.4
Gas jet apex angle			
45° (run C-68)	1.7	4.34	19.4
60° (run C-64)	1.7	4.14	20.6
Metal flow rate			
0.118 kg/sec (0.264 lb/sec) (run C82)	2.0	4.37	20.4
0.195 kg/sec (0.44 lb/sec) (run C64)	1.7	4.14	20.6
0.340 kg/sec (0.75 lb/sec) (run C80)	2.1	4.06	24.4
Metal superheat			
360°K (850°F) (run C62)	1.7	4.09	21.0
180°K (425°F) (run C64)	1.7	4.14	20.6
80°K (144°F) (run C63)	1.7	4.39	20.0

[a]Except as noted otherwise, all runs were made at the standard conditions of 4140 kN/m² pressure (600 psig), 73 mm (2.9 in) jet distance, 60° apex angle, and batch size of 7.5 ± 0.2 kg (16.5 ± 0.44 lb). In general, metal flow rate was 0.21 ± 0.02 kg/sec (0.46 ± 0.04 lb/sec), and metal superheat was 180 ± 10°K (425 ± 20°F).

and quench mechanisms, it has been possible to develop models for the prediction of size and shape in terms of atomization parameters. While further work is needed, it is believed that this work describes several important advances in the fundamental study of atomization during the past decade.

REFERENCES

Bradley, D. (1973). *J. Phys. D.* **6**, 1724.
Brunauer, S., Emmett, P. H., and Teller, E. (1938). *J. Amer. Chem. Soc.* **60**, 309.
Castleman, R. A. (1930). *Phys. Rev.* **35**, 1014.
Dombrowski, N., and Hooper, P. D. (1962). *Chem. Eng. Sci.* **17**, 291.
Dombrowski, N., and Johns, W. R. (1963). *Chem. Eng. Sci.* **18**, 203.
Elrod, H. G. (1954). *Heat. Piping Aircond.* March, 149–155.
Fraser, R. P., Eisenklam, P., Dombrowski, N., and Hassen, D. (1962). *AIChEJ* **8**, 672.
Grandzol, R. J. (1973). Water Atomization of 4620 Steel and Other Metals, Ph.D. dissertation, Dept. Chem. Eng., Drexel Univ., Philadelphia, Pennsylvania.
Grandzol, R. J., and Tallmadge, J. A. (1973). *AIChEJ* **19**, 1149.
Grandzol, R. J., and Tallmadge, J. A. (1975). *Int. J. Powder Met. Powder Tech.* **11**, 103.
Gummeson, P. U. (1972). *In* "Powder Metallurgy for High Performance Applications" (J. J. Burke and V. Weiss, eds.), Syracuse Univ. Press, Syracuse, New York.
Hirschhorn, J. S. (1969). "Introduction to Powder Metallurgy." Am. Powder Met. Inst., New York.
Kishidaka, H. (1972). *J. Soc. Powder Met.* **19**, 72.
Klar, E. and Shafer, W. M. (1972). *In* "Powder Metallurgy for High Performance Applications" (J. J. Burke and V. Weiss, eds.), Syracuse Univ. Press, Syracuse, New York.
Merrington, A. C., and Richardson, E. G. (1947). *Proc. Phys. Soc.* (*London*) **59**, Part I, 1.
Rao, P. (1970). M.S. thesis, Dept. Chem. Eng., Drexel University, Philadelphia, Pennsylvania.
Rao, P. (1973), Shape and Other Properties of Gas Atomized Metal Powders, Ph.D. dissertation, Dept. Chem. Eng., Drexel University, Philadelphia, Pennsylvania.
Rao, P., Grandzol, R. J., Schulz, N., and Tallmadge, J. A. (1970). *J. Vac. Sci. Technol.* **7**, No. 6, S 132.
Rao, P. and Tallmadge, J. A. (1972). *Proc. 1971 Fall Powder Met. Conf.*, (S. Mocarski, ed.). Metal Powder Industries Fed., New York.
Rayleigh, Lord (1878). *Proc. London Math. Soc.* **10**, 4.
See, J. B., Runkle, J. C., and King, T. B. (1973). *Met. Trans.* **4**, 2669.
Small, S., and Bruce, T. J. (1968). *Int. J. Powder Met.* **4**, (3) 7.
Squire, H. B. (1953). *British J. Appl. Phys.* **4**, 167.

Chapter 2

Triaxial Stress State Compaction of Powders

Robert M. Koerner

DEPARTMENT OF CIVIL ENGINEERING
DREXEL UNIVERSITY
PHILADELPHIA, PENNSYLVANIA

I. INTRODUCTION

The traditional method of forming powders into compacts has been the application of pressure. This pressurization can take many forms, as a recent compilation by Hausner (1967a) shows:

(a) unidirectional (die) pressing: single action pressing; double action pressing
(b) isostatic pressing
(c) powder rolling

(d) stepwise pressing
(e) powder extrusion: powder direct; powder canned
(f) powder swaging
(g) explosive compacting
(h) powder forging.

Despite the wide variety of techniques, no single technique is applicable to all situations. This is due primarily to variation in physical and chemical properties of the powders, variability of the final part geometry, and the desired ultimate part properties. These factors must be coupled with the selection of a particular compaction technique and must be within competitive and economic bounds. Thus certain applications will continue to require modifications to existing compaction procedures; e.g., rocking die compaction (Kotschy *et al.*, 1974), or totally new techniques, e.g., triaxial compaction. The concept, description, and performance of the triaxial compaction technique form the focal point of this chapter.

Triaxial compaction is a process which incorporates features of both unidirectional and isotatic compaction. Thus, both of these traditional methods will be examined to facilitate understanding of the new technique. Of greatest importance is an understanding of the global stress states in the different techniques. Through a firm understanding of shear and normal stresses, the benefits of triaxial compaction in achieving high density and strength can be realized.

First, the stress states are reviewed in unidrectional, isostatic, and triaxial compaction. Triaxial experimental apparatus and compaction procedures are then described. The overall density response and density gradients for a wide variety of powders are presented along with a comparison of transverse rupture strengths and density/strength responses for unidirectional, isostatic, and triaxial compaction methods. Implications of these results, with particular reference to production equipment conclude the chapter.

II. STRESS STATES IN POWDER COMPACTION

A. Unidirectional Compaction

In unidirectional, or die compaction, vertical pressure is applied to the powder, which in turn mobilizes lateral pressure from the die walls (Hausner, 1967b; Bockstiegel, 1966). As vertical deformation takes place, friction is developed between the powder and the die walls causing a nonuniform pressure distribution along the sides of the compact and a reduction in pressure along the base of the compact (Fig. 1). Information of this type has

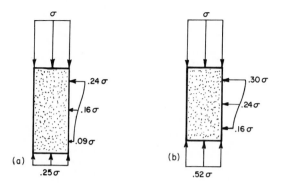

Fig. 1 Pressures mobilized during single action die compaction, (a) without and (b) with lubricant. (After Chukmasov and Zazimko, 1968).

given rise to double action pressing (to mobilize equal pressures at the top and bottom of the compact) and to the use of lubricants (to decrease die wall friction and thereby increase lateral pressure along the die walls). Even with these advances, however, the final global stress state is non-uniform. While production rates using this type of process are high, certain parts, particularly those with large height-to-diameter ratios and those of complex geometry cannot be compacted.

B. Isostatic Compaction

In isostatic, or hydrostatic, compaction the powder is placed in a flexible mold, immersed into a pressure vessel, and compacted via hydrostatic pressurization (Van Buren and Hirsch, 1966; Witkin, 1971). The global stresses are isostatic around the part being compacted, which allows for great flexibility in part size and part geometry. Unfortunately, production rates are very low, which has given rise to a modification of the original technique. This modification involves use of a fixed mold (or dry bag) instead of the flexible free mold (or wet bag). In dry-bag isostatic compaction, the upper and lower end closures are rigid while the sides are made of a flexible material. This is an important advancement because the powder can easily be filled in the mold, compacted, and ejected in a manner similar to die compaction. In fact, production rates can approach those of die compaction. One drawback is that the lateral pressure is transmitted through the consolidating powder to the upper and lower end closures, and the axial pressure is less than the lateral pressure. Figure 2 shows the approximate pressure states associated with both wet-bag and dry-bag isostatic compaction.

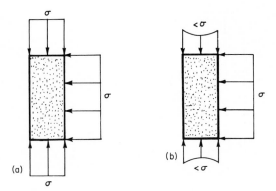

Fig. 2 Pressures mobilized during (a) free mold (wet-bag) and (b) fixed mold (dry-bag) isostatic compaction.

C. Triaxial Compaction

Triaxial compaction is a direct extension of the triaxial shear compression test used in geotechnical engineering to determine the shear strength of soils and rocks (see Lambe (1951) and Bishop and Henkel (1962)). Basically, the technique is an extension of dry-bag isostatic compaction. Using the same tooling as in fixed mold (or dry-bag) isostatic compaction, the upper and lower closures are loaded via an axial piston to pressures greater then the lateral pressure. Both lateral pressure and axial pressure are separately controlled giving a wide degree of freedom in pressure combinations. This global pressure state produces shear stresses within the compacting part which, as shown later, result in higher green density and strength than heretofore possible. It should also be mentioned that the basic technique of utilizing shear stresses during compaction has been treated by Buschow *et al.* (1968, 1969) in orienting particles for permanent magnets, and by Gregory (1959, 1960) in briquetting coal without a binder. In this latter work, torsional shear stresses were utilized with either the load piston or the tooling itself rotating. Thus, a previous concept of shear stress utilization in compaction is used here, except that the method of developing shear stresses is different and the range of stress states achievable is much larger.

For experimental study of triaxial compaction, a high-pressure cell capable of 100 ksi (690 MPa) was constructed (Fig. 3). In triaxial compaction, the powder to be compacted is placed in fixed mold (or dry-bag) tooling with rigid end platens. This assembly is placed in a pressure vessel and isostatic pressure is applied which eventually becomes the minor principal stress σ_3 (Fig. 3). At this point, the stress σ_3 acts in all directions, lateral as

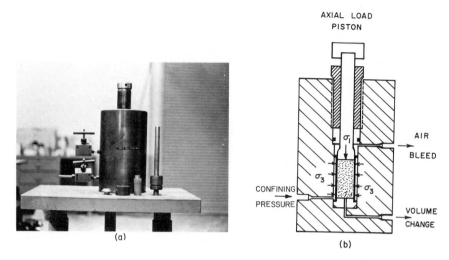

Fig. 3 (a) Photograph and (b) schematic of experimental triaxial compaction chamber used in this study.

well as axial. Axial pressure is then imposed by a piston to increase the axial stress to σ_1 while the lateral stress remains at σ_3. The resulting principal stress difference $\sigma_1 - \sigma_3$ is numerically equal to twice the shear stress τ.

If one views the process on Mohr's stress space of normal stress σ and shear stress τ, the response is as shown in Fig. 4(a). First, the confining pressure is increased to produce stress σ_3 in all directions, which is represented by the horizontal section of the dotted line. Next, the axial stress σ_1 is increased; this not only increases the average normal stress but produces a shear stress, as depicted by the angular section of the dashed line in Fig. 4(a).

It should be recognized that the sequencing of pressure application and the magnitudes of confining and axial pressures are important considerations. Stress paths of some of the various possible combinations are shown in Fig. 4(b). The stress path described in Fig. 4(a) is labeled ③. Other combinations range from ① (in which the confining and axial pressures are applied simultaneously) to ⑤ (where a large confining pressure is applied, then partially released, as axial pressure is applied). It was found that the total time for compaction was reduced using stress path ①, but the density of the resulting compacts was the lowest of all stress path routes investigated. Conversely, stress path ⑤ produced the highest density compacts, but the confining pressure utilized was greatest of the various alternatives attempted. For these reasons, stress path ③ is considered to be an optimum, and the remainder of the chapter will describe compacts formed by this sequence.

There is a limit to the amount of shear stress that can be applied to a com-

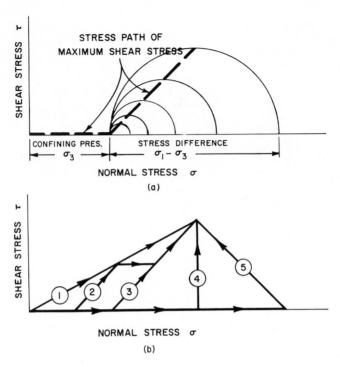

Fig. 4 Mohr stress space diagrams showing (a) standard triaxial stresses and (b) various alternative stress paths.

pact contained in a nonrigid mold, since shear failures occur. In this case, a failure plane propagates diagonally across the specimen at an approximate angle of $45° + \phi/2$, where ϕ is the angle of shearing resistance of the particulate mass of the compact. More discussion on this aspect of shear failure is treated in Terzaghi and Peck (1967). In order to determine this limit, iron powder was placed in flexible tooling, compacted to a given isostatic pressure, removed from the pressure vessel, and its dimensions measured. The green compact was then replaced in the pressure vessel and isostatic pressure returned to its original level. A standard dial deflection gauge and a volume change device (Bishop and Henkel, 1962), were set at initial values and the load piston was applied to the compact along its vertical axis. Readings of axial load, deflection, and volume change were made at frequent intervals. The calculated values of principal stress difference $\sigma_1 - \sigma_3$, axial strain, and volumetric strain $\Delta V/V$ are given in Fig. 5. Confining pressures of 15, 30, 50, 70, and 90 ksi (103, 207, 345, 483, and 621 MPa) were utilized.

The peak stress reached in each curve denotes the occurrence of shear failure. Increasing the confining pressure increases the principal stress dif-

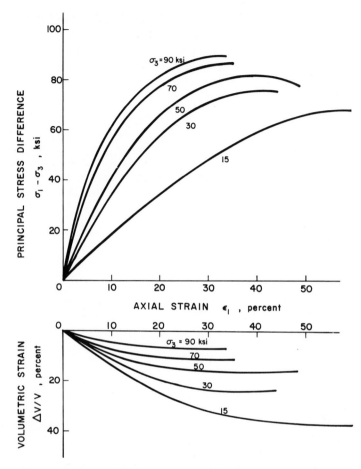

Fig. 5 Stress and volumetric strain versus axial strain response curves for green iron compacts at various confining pressures. (1 ksi = 6.9 MPa.)

ference, or shear stress, which can be sustained, i.e., the sample's shear strength is increased. The axial strain at which shear failure (maximum shear stress) occurs increases with decreasing confining pressure. The slope of the initial portion of the curves, i.e., the modulus of elasticity, increases with increasing confining pressure. This occurs because the green density also increases with increasing confining pressure. Volume decrease is most pronounced at the lower confining pressures and, thus, in the compacts of lower green density. The entire series of tests was performed on four different types of iron powders, but variation of results between the four were so small that only one set is presented. It can be concluded that, under shear stress, different iron powders behave similarily.

Having produced these curves at known values of confining pressure and determined their maximum principal stress difference, a Mohr's circle at failure can be drawn for each of the specimens tested. Figure 6 shows only the upper half of the circles because of symmetry. A curve is then drawn tangent to the circles and through the origin, giving the failure envelope for the material. The failure envelope represents the boundary above which stress combinations cause a shear failure. In this case, the curve was drawn through the origin since, at zero confining pressure, the uncompacted powder has no shear strength.

In the pressure range studied, the failure envelope terminates in a linear manner at an angle of approximately 5°. This value is consistent with results of Schwartz and Holland (1969), who tested iron powders at higher pressures and found a residual angle of 4°, and is similar to the approach taken by Afanesev *et al.* (1969). Empirically fitted on log–log paper, the equation of the failure curve is

$$\tau = 3.52 \, \sigma^{0.48}$$

Considering the slight difference in response between various iron powders tested and experimental error, it can be recommended that the shear strength of green iron compacts is given by

$$\tau = 3.5 \, \sigma^{0.5}$$

where τ is the shear strength strength (ksi) and σ is the normal stress (ksi).

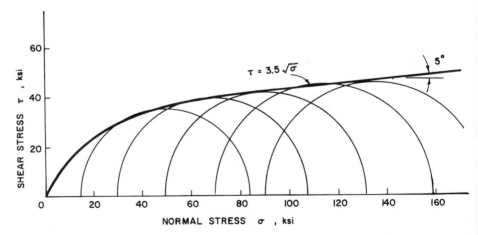

Fig. 6   Failure envelope on Mohr stress space for green iron compacts shown in Fig. 5. (1 ksi = 6.9 MPa.)

III. TRIAXIAL COMPACTION RESPONSE

A. Density

The basic density response in triaxial compaction is best represented as a plot of density versus shear stress at different confining pressures. This results in a family of curves starting with isostatic density, which corresponds to zero global shear stress, followed by the increase due to shear stress up to and including shear failure. Figure 7 gives the results for triaxial compaction of iron powder, but similar results have been obtained for other particulate systems listed in Table I (Koerner, 1971, 1973, 1975a,b, 1976; Koerner and McCabe, 1972a,b; Koerner and Quirus, 1971). Shear failure occurs when the curves reach a plateau, and the maximum density obtainable using triaxial compaction is found at the termination of the linear portion of these density response curves. This has been the general trend in behavior of all the

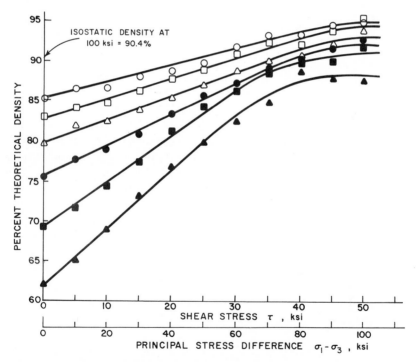

Fig. 7 Triaxial compaction response curves showing density increase under increasing shear stresses at various confining pressures. Confining pressure (ksi): ○, 60; □, 50; △, 40; ●, 30; ■, 20; ▲, 12. (1 ksi = 6.9 MPa.)

TABLE I

Summary of Powders Compacted Using the Triaxial Compaction Process

Powder		Approximate particle size (μm)	Particle shape	Relative response to cold triaxial compaction
Material	Type			
Iron	MH-100	100	Porous	Excellent
Iron	RZ-365 MM	100	Spherical	Excellent
Iron	ANCOR 1000	100	Irregular	Excellent
Iron	ATOMET 28	100	Irregular	Excellent
Aluminum	201 AB	70	Flake	Excellent
Copper	100 RXA	70	Flake	Excellent
Nickel	128 Carbonyl	8	Irregular	Excellent
Tungsten	C-20	7	Irregular	Excellent
Magnesia	MgO	3	Irregular	Excellent
Alumina	γ-Al$_2$O$_3$	0.05	Irregular	Good
Alumina	α-Al$_2$O$_3$	0.3	Irregular	Marginal
Uranium dioxide	UO$_2$	1000	Irregular	Good
Mock explosive	Ba(NO$_3$)$_2$	100	Irregular	Good
8620 steel	Gas atomized	300	Spherical	Poor
8620 steel	Annealed	300	Spherical	Poor
Tool steel	M-7	150	Spherical	None
Super alloy	11-B	70	Spherical	None
Charcoal	—	2000	Irregular	None
Coal	Bituminous	400	Irregular	Good

powders successfully compacted and listed in Table I. Additional details on specific response of a particular powder can be found in the references cited.

It is of interest to observe the generalized effectiveness of triaxial compaction on the various powders studied. Table II lists those metal and ceramic powders successfully compacted, arranged in order of increasing yield strength (for metals) or increasing hardness (for ceramics). Also shown is the amount of shear stress applied, which varies from isostatic, i.e., zero shear stress, to a relatively high value of 50 ksi (345 MPa). The table shows the average increase in percent theoretical density over the entire range of confining pressures. For example, iron powders increase in density 20% above apparent density during isostatic compaction, 36% with low shear stress, 46% with medium shear stress, and 51% with high shear stress. Thus, the density increase per unit shear is most pronounced at lower shear stress levels. To attain the higher densities often desired in industrial processes, higher shear stress is required. Also, the greater the hardness or yield stress of the powder, the less effective is the triaxial compaction technique in increasing density. This is understandable since particle deformation is required for large density increases.

TABLE II

Increase in Percent Theoretical Density above Apparent Density for Various Powders Compacted Triaxially with Increasing Levels of Shear Stress

Powder type	Shear stress in ksi (MPa)			
	0 Isostatic	10(69) Low shear	25(173) Medium shear	50(345) High shear
Metals[a]				
Aluminum, 17(117)	30[b]	44	47	47
Copper, 45(310)	23	42	54	60
Nickel, 105(725)	21	39	51	57
Iron, 120(828)	20	36	46	51
Tungsten, 220(1520)	12	22	30	34
Ceramics[c]				
Magnesia, 6	13	22	27	30
γ-Alumina, 8	10	19	24	26
α-Alumina, 9	—	4	6	7

[a]Yield stresses are shown in units of ksi (MPa).
[b]These are average values for tests at confining pressures ranging from 0 to 50 ksi (345 MPa).
[c]Moh hardness numbers are shown.

B. Density Gradients

In isostatic or triaxial compaction, it is generally considered that there is an absence of density variation in the compacted, or green, part because the applied pressures are uniform, relative to die compaction, Figs. 1 and 2. To determine if such statements are justified a comparative study was undertaken to measure density distribution. The powder used was an atomized iron powder (ANCOR 1000) compacted into right circular cylinders approximately 1 in. (25 mm) in diameter and 2 in (51 mm) high. The powders were vibrated in their tooling but not outgassed, and then compacted at either 30 ksi (207 MPa) or 60 ksi (414 MPa). The isostatically compacted samples were made by the wet-bag process. Triaxial samples had a medium amount of axial load applied, i.e., approximately equal to the confining pressure. The green compacts were removed from the tooling, cut into the desired sections, and their densities measured. Densities were determined by immersion in water after each sample was coated with a thin plastic spray to prevent air bubbles. Repeated tests showed that the measurements could be made to a precision of $\pm 0.4\%$.

Radial density variation was determined by first measuring the compact density, then machining off the outer $\frac{1}{8}$ in (3 mm), remeasuring the density, and continuing in this manner until only the center core remained. The results are shown in Fig. 8.

At 30 ksi (207 MPa) pressure the isostatic compact has higher density at the edge than in the center. The variation is about 0.23 g/cm³. This can be explained by considering relative particle movement within the compacting specimen. Particles closer to the edge move greater radial distances and thereby realign themselves into a denser packing configuration than do those particles at the center of the compact.

The triaxial compact at 30 ksi (207 MPa) confining pressure showed the same general tendency; however, the density variation was decreased to about 0.19 g/cm³. At 60 ksi (414 MPa) pressure, density variation of the isostatic compact from edge to center is decreased to approximately 0.10 g/cm³. The higher isostatic compacting pressures produce large particle deformation, which offsets the effect of particle movement discussed above. The triaxial compact at 60 ksi (414 MPa) confining pressure showed no density variation from the edge to the center of the compact.

Vertical density variation was obtained by sectioning the compacts horizontally and measuring each component part. Results are presented in

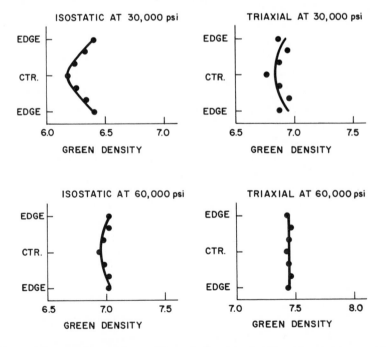

Fig. 8 Radial density gradients of isostatically and triaxially compacted iron powder at different levels of confining pressure. (1 ksi = 6.9 MPa.)

Fig. 9. At 30 ksi (207 MPa) isostatic compaction produced compacts having 0.2 g/cm³ density variation, with the higher density in the center and low density at the top and bottom. This is due to the relative rigidity of the tooling at the top and bottom of the compaction tooling, which produces the concave profile typical of isostatic compacts. At 30 ksi (207 MPa) confining pressure, the triaxial compacts show a significantly higher overall density and less variation from top to bottom although the general trend is the same as for the isostatic compact. At 60 ksi (414 MPa), the isostatic specimen showed a density variation of approximately 0.15 g/cm³ and varied from high density at midheight to low density at the top and bottom. Thus, high compaction pressures do not eliminate the low density regions at the top and bottom of the compact. Redesign of the tooling for greater flexibility at the top and bottom will reduce this effect. The triaxial compact at 60 ksi (414 MPa) confining pressure showed a density variation of only 0.03 g/cm³.

A series of compacts were formed at different height-to-diameter ratios ranging from 1.5 to 9.0 to determine the effect of this geometric factor on

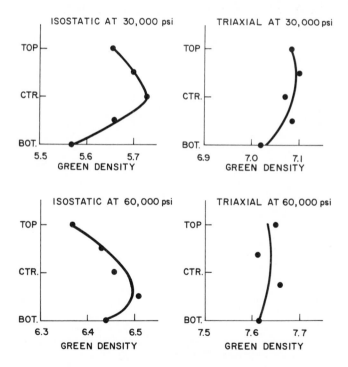

Fig. 9 Vertical density gradients of isostatically and triaxially compacted iron powder at different levels of confining pressure. (1 ksi = 6.9 MPa.)

vertical density variation. A compaction pressure of 30 ksi (207 MPa) was
used. At H/D of 1.5, a density variation was noted from top to bottom in both
isostatic and triaxial compacts, with the middle portion about 0.2 g/cm³
higher than the top and bottom. This compares well with the vertical density
gradient data previously described for compacts with H/D of 2.0. At H/D of
2.3, the isostatic compact had a density variation of approximately 0.1
g/cm³ with higher density in the center. At H/D ratios of 4.0 and above there
was no measurable density variation. The triaxial specimens at H/D of 2.3
and greater showed no density variation. Thus, the effect of the relatively
rigid ends of the tooling in producing low density sections at the top and
bottom of the compact is eliminated as H/D of the compact increases.

C. Strength

In order to determine the effect of shear stress on strength, some of the
sponge iron compacts were lightly sintered at 1000°F (538°C) for 1 hr in dry
hydrogen and then cut into rectangular bars. Transverse rupture test results
are shown in Fig. 10. Although there is considerable scatter in the results, it
is clear that, for a given density, the compacts formed under the highest shear
stress had the greatest strength. Note that, at 90% theoretical density,

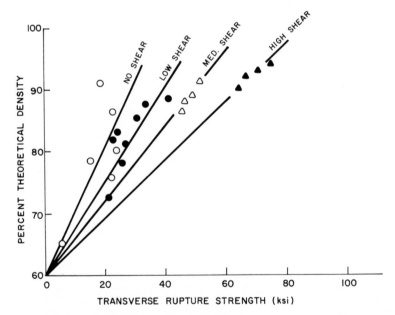

Fig. 10 Density versus transverse rupture strength for isostatically and triaxially compacted
iron powder compacts showing the influence of shear stress: ○, no shear (isostatic); ●, low
shear (0–17 ksi); △, medium shear (17–24 ksi); ▲, high shear (34–50 ksi). (1 ksi = 6.9 MPa.)

compacts formed by high shear stress had more than twice the strength of isostatic compacts, even though they were formed at significantly lower confining pressures. This suggests that shear stress during compaction not only increases density but has a secondary benefit of increasing compact strength for the same density.

IV. SUMMARY AND CONCLUSIONS

A summary of triaxial compaction is best presented through a comparison of results with die compaction and isostatic compaction. Table III illustrates this comparison for atomized iron powder (ANCOR 1000), first on the basis

TABLE III

Comparison of Green Density and Transverse Rupture Strength of Iron Compacts formed by Die, Isostatic, and Triaxial Compaction Methods[a]

(a) CELL PRESSURE COMPARISON			
COMPACTION METHOD	DIE	ISOSTATIC	TRIAXIAL
PRESSURES (ksi)	78, <78	60, 60	68, 12, 12
THEO. DENSITY	85 %	85 %	85 %
STRENGTH	20 ksi	25 ksi	55 ksi
(b) GREEN DENSITY COMPARISON			
COMPACTION METHOD	DIE	ISOSTATIC	TRIAXIAL
PRESSURES (ksi)	60, <60	60, 60	90, 60, 60
THEO. DENSITY	78 %	85 %	94 %
STRENGTH	15 ksi	25 ksi	73 ksi

[a] 1 ksi = 6.9 MPa.

of required cell pressure to reach an equivalent density, and second on the basis of densities resulting from equal cell pressures. In the first comparison, 85% of theoretical density can be reached by triaxial compaction in a cell capable of only 12 ksi (83 MPa), whereas the same density requires 78 ksi (538 MPa) in die compaction and 60 ksi (414 MPa) in isostatic compaction. Not only is less pressure required, but the strength is more than doubled in triaxial compaction over isostatic compaction. In the second comparison, 60 ksi (414 MPa) confining pressure is maintained in each compaction mode, and the density increase in triaxial compaction is 16% over die compaction and 9% over isostatic compaction.

In conclusion, triaxial compaction offers significant density and strength advantages over other compaction techniques because of the shear stresses developed. The main disadvantage, however, is the equipment requirement for addition of an independent axial pressure operation to what is essentially a fixed mold (or dry-bag) isostatic procedure. One suggestion for a production equipment scheme appears in Fig. 11. Two compaction chambers are interconnected on a reciprocating table which allows for compaction in one chamber while the other is ejecting the previously compacted part and then

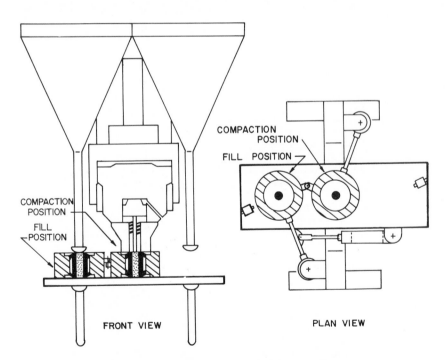

Fig. 11   Proposed high production triaxial compaction press.

being filled with new powder. Upon shifting chambers, the cycle is reversed. A second feature is that the same pressurizing system used for lateral confinement can be diverted and intensified to produce the axial pressure on the load piston. Although there are other systems which could be developed to accomplish triaxial compaction, the system shown has the economic benefits of low cycle time and high production rates.

Tooling life with triaxial compaction, like isostatic compaction, remains a practical production problem. With the scheme shown in Fig. 11, however, liners or knockouts can be used to minimize shutdown time for tooling changes.

Regarding part geometry, the triaxial compaction technique is limited to regular shapes, e.g., cylinders, squares, rectangles, etc., to ensure uniform pressure and densification. Success has also been realized on hollow cylinders, where the powder was laterally pressed against a solid metal core rod, and a matched annular piston was used to provide axial pressure. Elastic recovery of the compacted powder was sufficient to permit easy stripping of the compact from the core rod after compaction.

As noted in the introduction, no compaction technique is universally suited for all situations and there is always opportunity for innovation and development of new techniques. One such procedure is triaxial compaction, which can provide benefits of higher density and green strength, relative to die compaction and isostatic compaction, when used properly.

REFERENCES

Afanesev, L. N., Perelman, V. E., and Roman, O. V. (1969). *Porosh. Met. No. 6* (**78**), 34–39.
Bishop, A. W., and Henkel, D. J. (1962). "The Measurement of Soil Properties in the Triaxial Test," 2nd ed. E. Arnold, London.
Bockstiegel, G. (1966). *In* "Modern Developments in Powder Metallurgy" (H. Hausner, ed.), Vol. I, pp. 155–187. Plenum Press, New York.
Buschow, K. H. J., Luiten, W., Naastepad, P. A., and Westendorp, F. F. (1968). *Philips Tech. Rev.* **29**, 336–337.
Buschow, K. H. J., Naastepad, P. A., and Westendorp, F. F. (1969). *J. Appl. Phys.* **40**, 4029–4032.
Chukmasov, S. F., and Zazimko, A. I. (1968). *Porosh. Met. No. 6,* (**66**), 98–102.
Gregory, H. R. (1959). *Proc. Inst. for Briq. and Aggl., 6th Biennial Conf.,* pp. 38–49.
Gregory, H. R. (1960). *J. Inst. Fuel,* 447–461.
Hausner, H. H. (1967a). *In* "Perspectives in Powder Metallurgy" (H. H. Hausner, K. H. Roll, and P. K. Johnson, eds.), Vol. 1. Plenum Press, New York.
Hausner, H. H. (1967b). "Bibliography on the Compaction of Metal Powders." Hoeganaes Corp., Riverton, New Jersey.
Koerner, R. M. (1971). *Powder Met. Int.* **3**, 186–188.
Koerner, R. M. (1973). *Amer. Ceram. Soc. Bull.* **52**, 566–569.
Koerner, R. M. (1975a). *Proc. 2nd Int. Conf. on Compaction and Consolidation of Particulate Matter,* pp. 223–229.

Koerner, R. M. (1975b). *Proc. Int. Assoc. for Briquetting and Algomeration*, Vol. 14, pp. 9–19.

Koerner, R. M. (1976). *Proc. Int. Powder and Bulk Solids Handling and Processing Conf.*, in *Powder Bulk Solids Technol.* Fall, 1977, accepted for publication through Intl. Powder Inst., Chicago, Illinois 60690.

Koerner, R. M., and McCabe, W. M. (1972a). *Proc. 1971 Fall Powder Metallurgy Conf.* (S. Mocarski, ed.), pp. 267–279.

Koerner, R. M., and McCabe, W. M. (1972b). *Proc. 1972 Powder Metallurgy Conf.*, (A. S. Bufferd, ed.), pp. 225–241.

Koerner, R. M., and Quirus, F. J. (1971). *Int. J. Powder Met.* 7, 3–9.

Kotschy, J., Marciniak, A., and Sobczyk, M. (1974). *Int. Powder Met. Powder Tech.* 9, 135–137.

Lambe, T. W. (1951). "Soil Testing for Engineers." Wiley, New York.

Schwartz, E. G., and Holland A. R. (1969). *Int. J. Powder M et.* 5, 79–87.

Terzaghi, K., and Peck, R. B. (1967). "Soil Mechanics in Engineering Practice." Wiley, New York.

Van Buren, C. E., and Hirsch, H. H. (1966). *In* "New Types of Compacting Methods," pp. 27–65. Plenum Press, New York.

Witkin, D. E. (1971). "Isostatic Pressing: A Review of the State of the Art." American Ceramic Society, Annual Meeting, Washington, D.C., May 1972.

Chapter 3

Diffusional Homogenization of Compacted Blends of Powders

Richard W. Heckel

DEPARTMENT OF METALLURGICAL ENGINEERING
MICHIGAN TECHNOLOGICAL UNIVERSITY
HOUGHTON, MICHIGAN

I. INTRODUCTION

In conventional processing of powders, the compacted powder form is sintered, i.e., subjected to a high temperature environment with a protective atmosphere. The elevated temperature accelerates atomic diffusion and,

51

by various mechanisms, bonds adjoining particles to each other, thus developing strength in the compacted powder mass. Much has been written on the mechanisms of bonding during sintering and they will not be treated here. Rather, this chapter focuses on the use of diffusional transport during sintering for development of alloys from elemental powder particles.

Alloys may be produced by powder metallurgy techniques in a wide variety of ways. The most common method is the use of *prealloyed powders* wherein all of the particles have the same composition, thus assuring a homogeneous alloy throughout all of the processing steps. Prealloyed powders are typically produced by atomization of a stream of the molten alloy, as described in Chapter 1. Other methods of producing alloys include:

(1) *Mechanical alloying*, wherein constituent powders are blended and deformed by a process such as ball milling to achieve essentially homogeneous powder particles;

(2) *Infiltration*, wherein a liquid of one composition is caused to penetrate the pores of a compact of a different composition. Copper infiltration of an iron compact is a common example and results in a composite of iron and copper with a small amount of alloying;

(3) *Reduction of finely divided oxide particles* to achieve an essentially homogeneous alloy powder. This method has also been used to produce powders with fine, dispersed oxides in a metal or alloy matrix for fabrication of dispersion strengthened materials;

(4) *Homogenization processing of alloys*, wherein constituent powders are blended, compacted and given a subsequent thermal treatment to achieve the required degree of homogeneity. Typically, the homogenization process takes place by solid-state interdiffusion of the elements in the blend. The thermal process may, however, result in the formation of a liquid phase, as in the case of blended Cu and Sn powders (to produce bronze bearings).

There are a number of advantages to homogenization processing.

(a) Alloys can be produced from readily available powders. Many of the elements of the periodic table are available in the form of high-purity powders. Compositional adjustments can be readily achieved by varying the proportions of powders in the blends. Furthermore, the tramp element impurity problems inherent in powders made by melting and atomizing can be minimized by using high-purity elemental powders and homogenization processing (Heckel and Erich, 1972).

(b) Higher green strengths in compacts occur. If elemental metal powders are used, their high ductility will result in high-density compacts and, consequently, high green strength. A blend of an elemental powder (the base element of the alloy to be formed) and a master alloy powder (containing all of the alloying elements) can often be used to achieve high green strengths for highly alloyed materials (Koczak, 1976).

(c) There is control of microstructure. Homogenization processing offers the potential for producing partially homogenized structures or composites.

(d) There is overall process economy. Many metals are manufactured by chemical processes which directly yield a powder without the high temperature (and cost) necessary to produce a molten metal. Thus, there are potential economic (and often environmental) advantages in eliminating pyrometallurgical processes in the fabrication of alloys.

It should also be pointed out that distinct technological advantages in terms of improved alloy performance can be achieved relative to cast and wrought processing through homogenization processing of alloys. For example, fully dense Hastelloy B sheet produced by powder blending (Ni, Mo, and Fe elemental powders), compaction, sintering, and alternate cold rolling and thermal treatment has been shown to have improved corrosion resistance and essentially equivalent mechanical properties when compared to cast and wrought Hastelloy B (Heckel and Erich, 1972).

The disadvantages to homogenization processing mainly center about the additional processing steps that are needed and the necessity for suitably characterizing the development of a homogeneous composition in the alloy. The additional steps include blending the constituent powder particles and thermally treating the compacted blend to remove compositional heterogeneities. Characterization of the state of homogeneity can be accomplished by a wide variety of direct and indirect measurements (Heckel *et al.*, 1970). Qualitative, and sometimes quantitative, microscopy is of value in this regard, along with electron probe microanalysis, energy dispersive X-ray analysis with a scanning electron microscope, and X-ray compositional line broadening. An additional disadvantage often arises when the alloy of interest contains a readily oxidized element such as Cr. In such instances care must be taken to use highly reducing atmospheres, especially when interconnected porosity is present, and to take precautions to assure that very low amounts of oxygen are present in the powders.

It is thus clear that powder fabrication of alloys by diffusional homogenization of compacted blends of powders requires the careful control of a number of factors which are not germane to more conventional powder fabrication (e.g., alloys via the use of prealloyed powders). These factors concern the necessity to design processing routes which will achieve desired levels of homogeneity without resorting to extremes that are economically unattractive. (For example, the specification of inordinately long duration and/or high-temperature thermal treatments, or very fine particle sizes, would make homogenization processing unattractive.) It is, therefore, important to obtain a quantitative understanding of the magnitude of the effects of various factors on homogenization kinetics.

II. HOMOGENIZATION VARIABLES

The large number of variables that can affect the kinetics of homogenization can be classified as powder parameters, processing parameters, and alloy system parameters. These will be discussed qualitatively at this point; quantitative treatments will be provided in a later section.

A. Powder Parameters

1. Particle Size

Homogenization kinetics can be related directly to the square of the diffusion distance (Rhines and Colton, 1942). Powder particle size determines the diffusion distance and, therefore, plays a significant role in homogenization processing. However, it should be noted that the overall geometry of the compacted blend must be considered. In instances where two types of powder are blended, one type of particle will generally be dispersed in a rather continuous matrix composed of particles of the other type. Such situations result in the particle size of the dispersed powder (generally the minor constituent of the blend) having the primary effect, since, for the most part, the particulate nature of the other powder has been lost in the compaction process.

2. Particle Shape

Generally, shape is not important in the overall kinetics of the process since diffusion distances are usually long relative to the particle dimensions. Actually, it is the shape of the diffusion field that is critical; only extreme variations in particle shape will distort the diffusion field significantly from a spherical symmetry.

3. Particle Composition

Particle composition can affect the homogenization process in many ways. Certainly, impurities enter the compact, in part, via the powders; undesirable elements should, therefore, be minimized. Particle compositions in the blend naturally define the alloy system; these parameters will be considered below. Particle compositions also have an effect on the degree of homogenization that is necessary, since the initial magnitude of compositional heterogeneity is defined by the particle compositions.

4. Particle Structure

This parameter affects homogenization in several ways. Hard particles (e.g., cold worked) are difficult to deform during the compaction process and can, therefore, result in different microstructures than would be formed using soft, annealed particles. Oxide layers on particle surfaces can form diffusion barriers which impede the interdiffusion process. Even readily reduced oxides are generally undesirable since they can be trapped between particles and, therefore, become inaccessible to reducing gases in the sintering atmosphere. Furthermore, even if the oxides are reduced by the sintering atmosphere, the products formed by the reduction process could react with the other powder constituent of the blend, if the formation of a highly stable oxide is possible. Coated powder particles (composite microstructure) have been studied from a homogenization processing standpoint by Lund *et al.* (1960). A particle of a given composition is coated with another metal rather than blending two different powders together. Obviously, this process eliminates the blending step in homogenization processing and thereby precludes the possibility of slow kinetics due to poor interparticle mixing.

B. Processing Parameters

1. Blending (Mixing)

Blending of the constituent powders is a critical step in homogenization processing; poorly blended powders will homogenize more slowly than anticipated since interdiffusion distances can become large. Blending can be difficult for constituent powders which have significantly different densities, particle sizes, and sometimes surface characteristics and shapes. It is sometimes possible to achieve satisfactory blends by using these effects to offset each other (e.g., by blending coarse particles of the higher density powder with fine particles of the lower density powder). However, it is often necessary to add a very small amount of volatile liquid to the powder mix to minimize gravity segregation; the volatile liquid can be allowed to evaporate from the compact prior to or during subsequent thermal treatment.

2. Ratio of Powders in the Blend (Mean Composition of the Compact)

The effects of compact composition on homogenization behavior can be complex and it is, therefore, difficult to consider this parameter alone. However, it is easily seen that decreasing the number of minor constituent particles in a blend (i.e., decreasing the mean solute content) increases separation between the particles by increasing the amount of solvent-rich matrix. Thus, in general, homogenization kinetics are decreased with decreasing mean solute concentration.

3. Thermal Processing Conditions (Temperature Effect)

High-temperature homogenization of compacted blends of powders takes place predominantly by volume (or lattice) interdiffusion.† Thus, the primary effect of temperature on process kinetics is in the interdiffusion coefficient (or coefficients, in the case of a multiphase homogenization situation)

$$\tilde{D}_i = D_{0i}\exp(-Q_i/RT) \qquad \text{(for the } i\text{th phase)} \qquad (1)$$

Where $\tilde{D}_i$ is the interdiffusion coefficient, D_{0i} the preexponential coefficient, Q_i the activation energy, R the universal gas constant, and T the absolute temperature. For multiphase situations, temperature also affects the homogenization kinetics through changes in phase solubilities.

4. Mechanical Working Conditions

Two different kinds of mechanical working can be considered. The application of pressure during the powder compaction process to achieve a given compact density, and the deformation processing of sintered compacts to achieve further increases in density. Increased compaction pressure results in higher compact densities. However, the effect of compaction pressure on homogenization kinetics is complex. Higher densities result in better interparticle contact which should improve interdiffusion between the particles. On the other hand, lower densities result in a higher surface area in the compact and thus a larger contribution of surface diffusion to homogenization. A complete understanding of the effects of compaction pressure (or compact density) on homogenization kinetics is not available; a research program on this topic would be very valuable.

---

† As will be shown later, the contributions of other transport mechanisms can be observed as temperature is decreased. Low-temperature kinetics are often found to be greater than predictions based on volume diffusion, presumably due to surface and grain boundary diffusion.

Mechanical working after sintering, but prior to a specifically designed thermal treatment to effect homogenization, must be considered from two standpoints:

(i) introduction of lattice defects which would increase the interdiffusion coefficient; and

(ii) alteration of the dimensions of the particles in the compact as well as the overall geometry of the microstructure of the compact.

Lattice defects will anneal out very rapidly if high-temperature homogenization treatments are used; thus, they probably contribute little to increasing homogenization kinetics if normal heat treating procedures are employed. Geometry changes during the mechanical working of sintered compacts are best understood by considering a dispersion of spherical particles of component B in a matrix of A. If the component B particles deform along with A matrix, the spherical geometry is changed to a cylindrical one by extrusion or to a laminar (or planar) one by rolling. The decrease in cylinder radius or plate thickness by mechanical working obviously decreases the interdiffusion distance and thus speeds homogenization. On the other hand, the change in geometry of the diffusional fluxes from spherical (three-dimensional) to cylindrical (two-dimensional) or planar (one-dimensional) slows the process. It appears that the effect of reduction in particle dimension and the effect of flux geometry change offset each other for small amounts of deformation; large amounts of deformation cause the particle dimension effect to dominate and result in an increase in homogenization kinetics.

C. Alloy System Parameters

1. Microstructure (Phase Solubilities)

For binary alloys, the appropriate equilibrium phase diagram is of value as a guide to the phases that will form in the microstructure during the homogenization process. Since homogenization treatments are typically carried out at high temperatures, the composition ranges of phases present in the microstructure correspond closely with those present on the phase diagram. For higher order systems the interdiffusion path is not uniquely defined by the isotherm corresponding to the homogenization temperature because of the additional degree(s) of compositional freedom. The additional complexities associated with homogenization in ternary and higher order diffusion systems are discussed by Purdy and Kirkaldy (1971).

The nature of interdiffusion processes occurring during homogenization is also defined by the alloy system and the phases present in the microstructure. For binary systems, only one-phase fields form during inter-diffusion; these phase fields have associated with them specific interdiffusion coefficients which, in general, are composition dependent as well as temperature dependent, Eq. (1). In many instances though, accurate descriptions of the interdiffusion process can be achieved by using a single interdiffusion coefficient representing the coefficient for each phase in the microstructure.

III. ANALYSIS OF VARIABLES

It may be inferred from the preceding section that wide variety of para-meters can be important in determining the kinetics of homogenization of compacted blends of powders. Many of the variables are interrelated, mak-ing it difficult to deal with them individually. However, it is possible to deal with them in groups by the development of mathematical models which describe the homogenization process. Such models are of value in defining the relative effects of many of the parameters which have been discussed as well in predicting the overall homogenization kinetics. Since the models involve simplifying assumptions, a later section of this chapter will evaluate the accuracy of the models.

A. Microstructures

It was pointed out previously that the phase diagram defines the phases present during the homogenization process for binary alloys. The additional information needed for this definition are the mean solute concentration of the alloy $\bar{C}$, the solute concentrations in each of the two powders of the blend, and the homogenization temperature.

1. One-Phase Binary System (Cu–Ni)

The Cu–Ni phase diagram exhibits complete solid solubility of Cu and Ni (Hansen, 1958). The homogenization process, therefore, proceeds simply by leveling of the concentration–distance profiles in the compact, see Heckel (1964). Figure 1 shows a sequence of microstructures developed during the homogenization of compacts having a mean composition of 80 at. % Ni and 20 at. % Cu ($\bar{C} = 0.20$ a./f. Cu).† Figure 1(a) shows the as compacted structure; Fig. 1(b) shows the partially homogenized structure

---

† a.f. refers to atomic fraction.

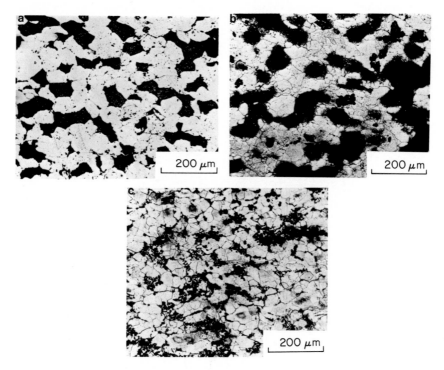

Fig. 1 Microstructures of blends of Cu and Ni powders ($\overline{C} = 0.20$ a.f. Cu) compacted at 557 MPa (80,000 psi). (a) As compacted, (b) thermally treated at 1050°C (1920°F) for 45 min. ($F = 0.55$), and (c) thermally treated at 1050°C (1920°F) for 10 hr ($F = 0.84$). (Heckel and Balasubramaniam, 1971.)

(characterized by a diffuse etching pattern); Fig. 1(c) shows the almost completely homogenized structure. Note in Fig. 1(c) that Kirkendall porosity (Smigelskas and Kirkendall, 1947), resulting from the fact that the Cu atoms have a higher intrinsic diffusivity than the Ni atoms, has developed as expected at the original locations of the Cu particles.

2. Two-Phase Binary System (Ni–W)

The Ni–W phase diagram in the vicinity of 1200°C (2200°F) consists of the face-centered cubic (fcc) Ni-rich and body-centered cubic (bcc) W-rich terminal solid solutions separated by a two-phase field Hansen (1958). Blends of Ni and W powders with mean compositions in the two-phase field would undergo interdiffusion during homogenization (i.e., the Ni would dissolve W, and the W would dissolve Ni), but two phases would always exist. If the mean composition were in the Ni-rich fcc field, the W-rich phase would dissolve completely with subsequent homogenization taking place by

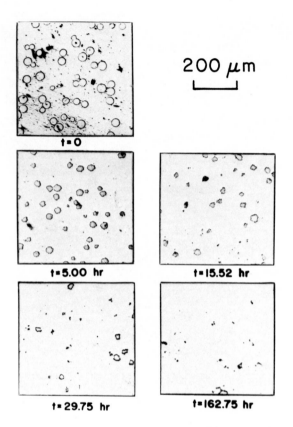

Fig. 2 Microstructures of blends of Ni and W powders ($\overline{C}$ = 0.107 a.f. W) compacted at 679 MPa (98,400 psi) and thermally treated for various lengths of time at 1156°C (2110°F). W 325/400 mesh. Ni–400 mesh. (Tanzilli and Heckel, 1975.)

leveling of the concentration gradients in the Ni-rich phase. The first stage of homogenization of Ni-W compacts with $\overline{C}$ = 0.107 a.f. W (in the Ni-rich field at 1156°C (2110°F) is shown in Fig. 2. An indication of the concentration–distance profiles around the dissolving W particles is shown by the photomicrograph in Fig. 3. There are obvious similarities between the homogenization of two-phase systems and the solution treatment of cast and wrought two-phase alloys, as shown by Tanzilli and Heckel (1975) and Baty *et al.* (1970).

3.   Three-Phase System (Ni–Mo)

The Ni–Mo phase diagram in the vicinity of 1200°C (2200°F) consists of the fcc Ni-rich (α) and bcc Mo-rich (γ) terminal solid solutions and the MoNi intermediate phase (β) (Hansen, 1958). Various possible homogeniza-

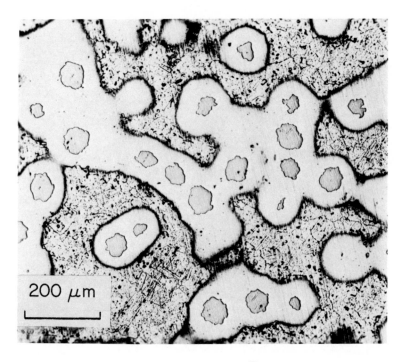

Fig. 3 Microstructure of a Ni–W powder compact ($\overline{C} = 0.107$ a.f. W) thermally treated at 1207°C (2205°F) for 20.5 hr. Gray W-rich β-phase particles are surrounded by the α-phase matrix; unetched α-phase regions show the location of high W concentrations ($\sim0.05 < C_W < C_\beta$). (Tanzilli and Heckel, 1975.)

tion sequences can take place in this system depending on the powder particle and mean compact compositions. If pure Ni and Mo powders are blended and compacted, the initial stage of the homogenization process will be formation of layers of the MoNi (β) phase at the interface between the dispersed Mo particles and the Ni matrix (Fig. 4) (Lanam *et al.*, 1975). Subsequent stages depend upon the mean compact composition. If the mean composition lies in the Ni-rich solid solution field, the Mo-rich phase will dissolve, creating a two-phase homogenization situation which will behave in a manner similar to the Ni–W situation described above.

4. Microstructural Model

Typically, compacted blends of powders exhibit microstructures which consist of isolated particles of the minor constituent powder in a matrix of the major constituent. The interparticle porosity exists mainly in the matrix with some pores located at the interfaces between particles of

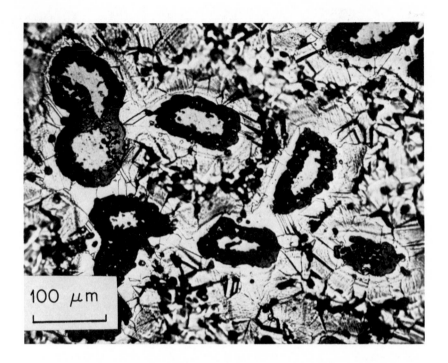

Fig. 4 Microstructure of a Ni–Mo powder compact after a short thermal treatment. A dark etching layer of MoNi (β) has formed at the original interface between the dispersed Mo particles and the Ni matrix. (Lanam and Heckel, 1975.)

different type. Dispersion of the minor constituent is improved by decreasing the particle size of the major constituent, leading to enhanced homogenization kinetics. (Two minor constituent particles which are contiguous behave as a single particle of larger size.)

The concentric-sphere geometry has been proposed by Raichenko (1961) and Fisher and Rudman (1961) to provide a basis for the formulation of mathematical models of homogenization kinetics. This idealized geometry consists of a spherical particle of the minor constituent surrounded by a shell of the major constituent of the blend, with the mean solute composition of the sphere–shell composite being the same as that of the entire compact,

$$\overline{C} = (l/L)^3 \tag{2}$$

where l is the diameter of the minor constituent sphere (minor constituent particle size) and L is the diameter of the composite. The sphere–shell symmetry element is shown in Fig. 5 along with the planar and cylindrical elements which will be used in developing models that apply to microstructures produced by compact rolling and extrusion, respectively.

The concentric-sphere geometry, although it provides a reasonable description of the microstructure, has several shortcomings. Certainly, this geometry assumes spherical minor constituent particles, a perfect distribution of these particles, and perfect contact at the sphere–shell interface. Furthermore, it assumes space filling by the spherical composites (which is impossible)† and it ignores any interparticle porosity. (It is noteworthy that the particle size of the major constituent is excluded from the geometric model, since the major constituent is assumed to be formed into a continuous matrix by the compaction process.) Nevertheless, the concentric-sphere model has been shown in various studies to provide a reasonable basis for constructing mathematical models to predict homogenization kinetics and for understanding the interdependence of the various parameters associated with the interdiffusion process. Other models have also been proposed, but these are based on alternate-plate (planar) and ordered-cube geometries and do not have the general flexibility and applicability of

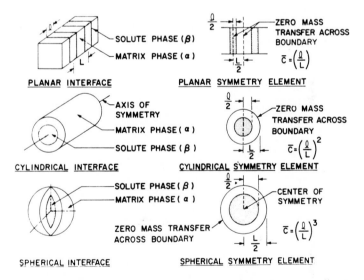

Fig. 5 Planar, cylindrical, and spherical finite geometries and corresponding symmetry elements. (Tanzilli and Heckel, 1968.)

---

†Another way of considering this assumption is to recognize that the boundary condition which will be applied at $r = L$ is a concentration gradient, $\partial C/\partial x$, of zero. In a real situation, the locus of $\partial C/\partial x = 0$ will not be spherical.

the concentric-sphere model (Chevenard and Wache, 1944; Duwez and Jordan, 1949; Weinbaum, 1948; Gertziken and Feingold, 1940; Raichenko and Fedorchenko, 1958).

An understanding of the nature of the solute concentration profiles as a function of radial distance in the sphere–shell composites is necessary for the formulation of mathematical models for binary systems. Schematic concentration–distance profiles for one-, two-, and three-phase systems are shown in Fig. 6. C_{ij} is defined as the solute concentration in the ith phase at the i–j interphase interface, ξ_i is the diameter of the spherical interphase interface surrounding the ith phase, and T_H is the homogenization temperature. Schematic drawings of the microstructures in Fig. 6 can be compared to the actual microstructures of one-, two-, and three-phase systems shown in Figs. 1–4.

B. Kinetic Models

In general, the formulation of mathematical models to describe interdiffusion processes involves the simultaneous solution of Fick's law (one equation for each phase) and the flux balance equation (one for each interface) for the appropriate initial and boundary conditions and geometry. For the concentric-sphere model, Fick's law for concentration-independent interdiffusion coefficients is

$$\partial C/\partial t = \tilde{D}_i[\partial^2 C/\partial r^2 + (2/r)\partial C/\partial r] \qquad \text{(spherical)} \qquad (3)$$

where C is the solute concentration, r the radial distance from the center of the sphere, and t is time. Planar and cylindrical equivalents of Eq. (3) are

$$\partial C/\partial t = \tilde{D}_i \partial^2 C/\partial x^2 \qquad \text{(planar)} \qquad (4)$$

and

$$\partial C/\partial t = \tilde{D}_i[\partial^2 C/\partial r^2 + (1/r)\partial C/\partial r] \qquad \text{(cylindrical)} \qquad (5)$$

The flux balance equation which describes the interphase interface velocity is

$$(C_{ji} - C_{ij})d(\xi_{ij}/2)/dt = \tilde{D}_i[\partial C_i/\partial r]_{\xi_{ij}/2} - \tilde{D}_j[\partial C_j/\partial r]_{\xi_{ij}/2} \qquad (6)$$

where the solute concentration difference on the left is the concentration discontinuity at the interface, $d(\xi_{ij}/2)/dt$ is the interface velocity, and the two terms on the right are the interdiffusion fluxes in the ith and j^{th} phases at the interface. Equation (6) is valid for

$$\overline{V}_i = \overline{V}_j = \text{const} \qquad (7)$$

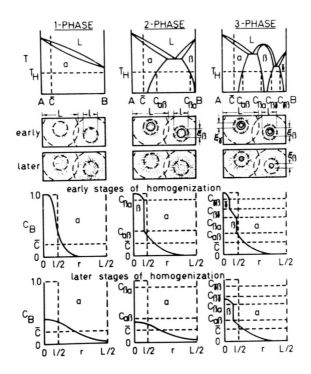

Fig. 6 Schematic representation of the homogenization process in one-, two-, and three-phase binary systems in terms of microstructural changes and composition–distance profiles.

for both components of the alloy (where $\overline{V}_i$ and $\overline{V}_j$ are the partial molar volumes in the ith and j^{th} phases) (Sekerka *et al.*, 1975). Furthermore, Eq. (6) is applicable to planar, cylindrical, and spherical geometries. An example of the form of Eq. (6) for a specific interface, the α–β interface in the two-phase situation defined in Fig. 6, would be

$$(C_{\beta\alpha} - C_{\alpha\beta})\, d(\xi_\beta/2)/dt = \tilde{D}_\alpha[\partial C_\alpha/\partial r]_{\xi_\beta/2} - \tilde{D}_\beta[\partial C_\beta/\partial r]_{\xi_\beta/2} \qquad (8)$$

The initial and boundary conditions for the concentric-sphere homogenization model for situations where pure elemental powders are blended and compacted are

initial conditions: $C = 1$ a.f. solute, $0 \leq r < l/2$
                        $C = 0$ a.f. solute, $l/2 \leq r \leq L/2$
boundary conditions: $\partial C/\partial r = 0$, $r = 0$, $r = L/2$
                        C_{ij}(and C_{ji}) = const, $r = \xi_{ij}/2$.

Appropriate changes must be made in the initial conditions if binary alloy powders are used. Furthermore, if the composition of one or both of the constituent powders would be in a two-phase field, it would be necessary to alter Eq. (6) to reflect the loss of flux on the two-phase side of the interface and the change in magnitude of the compositional discontinuity at the interface (Wagner, 1960; Sekerka *et al.*, 1975).

The inherent complexity in diffusional homogenization models, since they necessarily involve finite geometry situations, generally dictates the use of numerical methods and computer techniques to carry out the necessary computations. This is especially the case for multiphase problems. The grid techniques used extensively for heat conduction problems are described in numerous texts (for example, see Carslaw and Jaeger, 1959) and are readily adapted to diffusional problems (Sekerka *et al.*, 1975). It is usually helpful in treating multiphase problems to use a variable grid technique in order to have an integral number of grids in each phase while accommodating the moving interface(s) (Murray and Landis, 1959).

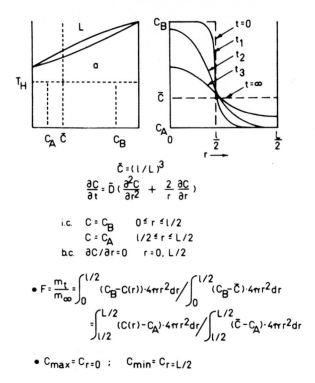

$$\bar{C} = (l/L)^3$$

$$\frac{\partial C}{\partial t} = \bar{D}\left(\frac{\partial^2 C}{\partial r^2} + \frac{2}{r}\frac{\partial C}{\partial r}\right)$$

i.c. $\quad C = C_B \qquad 0 \leq r \leq l/2$

$\qquad C = C_A \qquad l/2 \leq r \leq L/2$

b.c. $\quad \partial C/\partial r = 0 \qquad r = 0, L/2$

$$\bullet \; F = \frac{m_t}{m_\infty} = \int_0^{l/2}(C_B - C(r))\cdot 4\pi r^2 dr \bigg/ \int_0^{l/2}(C_B - \bar{C})\cdot 4\pi r^2 dr$$

$$= \int_{l/2}^{L/2}(C(r) - C_A)\cdot 4\pi r^2 dr \bigg/ \int_{l/2}^{L/2}(\bar{C} - C_A)\cdot 4\pi r^2 dr$$

$$\bullet \; C_{max} = C_{r=0} \; ; \quad C_{min} = C_{r=L/2}$$

Fig. 7 One-phase, concentric-sphere homogenization model. (Heckel and Balasubramaniam, 1971.)

1. One-Phase Model

Formulation of the one-phase concentric-sphere model is defined in Fig. 7 for the situation where the initial powder particle compositions are C_A and C_B (Heckel and Balasubramaniam, 1971). Also shown in Fig. 7 are measures by which the progress of homogenization can be characterized in one-phase systems:

(i) the degree of interdiffusion F, defined as the ratio of the mass transferred through the surface $r = l/2$ in time t to that which will be transferred in infinite time ($0 \leq F \leq 1$);

(ii) the compositions existing at positions $r = 0$ and $r = L/2$ which define the range of compositions existing at any time ($\overline{C} < C_{max} \leq C_B$ and $C_A \leq C_{min} < \overline{C}$).

Calculated values of F as a function of mean compact composition $\overline{C}$ (using the expressions in Fig. 5), geometry [using Eqs. (3)–(5)], and reduced time $\check{D}t/l^2$, are given in Fig. 8. The maximum and minimum solute concentrations are shown in Fig. 9 for the same variables.

It may be noted that the geometry and mean composition dependence of F in Fig. 8 results primarily in a lateral shift in the various curves. Thus, it is possible to define empirically a new reduced time parameter $(\check{D}t/l^2)(n^2 Y)$,

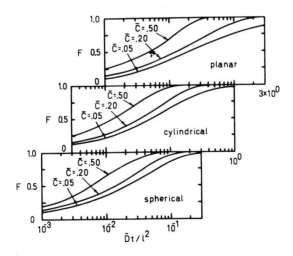

Fig. 8 Degree of interdiffusion F calculated by the one-phase planar, cylindrical, and spherical models as a function of reduced time $(\check{D}t/l^2)$ for various mean compositions $(\overline{C})$. (Heckel and Balasubramaniam, 1971.)

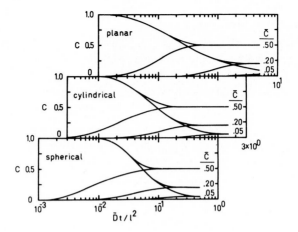

Fig. 9 Range of compositions (C_{min} and C_{max}) calculated by the one-phase planar, cylindrical, and spherical models as a function of reduced time ($\check{D}t/l^2$) for various mean compositions ($\overline{C}$). (Heckel and Balasubramaniam, 1971.)

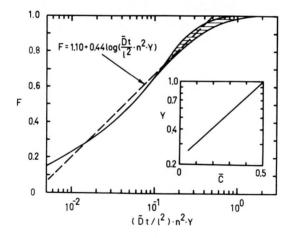

Fig. 10 Normalized degree of interdiffusion (F) as a function of ($\check{D}t/l^2$). $n^2 Y$, which collapses the F versus t curves (Fig. 8) to a single curve. Planar, $n = 1$; cylindrical, $n = 2$; spherical, $n = 3$. (Heckel and Balasubramaniam, 1971.)

which allows for the graphical expression of the F curves in Fig. 8 as a single curve (Fig. 10) with only a minor amount of scatter at the larger values of F. Figure 10 provides a means for evaluationg the effects of homogenizing temperature (through $\check{D}$, see Eq. (1)), minor constituent particle size l, mean

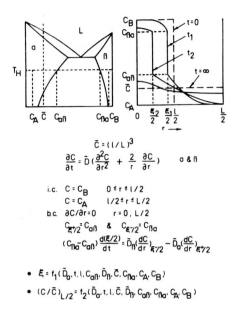

$$\bar{C} = (l/L)^3$$

$$\frac{\partial C}{\partial t} = \bar{D}\left(\frac{\partial^2 C}{\partial r^2} + \frac{2}{r}\frac{\partial C}{\partial r}\right) \quad \alpha \,\& \, \beta$$

i.c. $\quad C = C_B \qquad 0 \leq r \leq l/2$
$\qquad\quad C = C_A \qquad l/2 \leq r \leq L/2$

b.c. $\quad \partial C/\partial r = 0 \qquad r = 0, L/2$
$\qquad\quad C_{\xi/2} = C_{\alpha\beta} \quad\&\quad C_{\xi/2} = C_{\beta\alpha}$

$\qquad\quad (C_{\beta\alpha} - C_{\alpha\beta})\dfrac{d(\xi/2)}{dt} = \bar{D}_\beta\left(\dfrac{dC}{dr}\right)_{\xi/2} - \bar{D}_\alpha\left(\dfrac{dC}{dr}\right)_{\xi/2}$

- $\xi = f_1(\bar{D}_\alpha, t, l, C_{\alpha\beta}, \bar{D}_\beta, \bar{C}, C_{\beta\alpha}, C_A, C_B)$
- $(C/\bar{C})_{L/2} = f_2(\bar{D}_\alpha, t, l, \bar{C}, \bar{D}_\beta, C_{\alpha\beta}, C_{\beta\alpha}, C_A, C_B)$

Fig. 11 Two-phase, concentric-sphere homogenization model. (Heckel and Balasubramaniam, 1971.)

composition $\bar{C}$, and geometry (through n) on the progress of homogenization in one-phase systems.

2. Two-Phase Model

Formulation of the two-phase, concentric-sphere model is defined in Fig. 11 for the situation where the initial powder particle compositions are C_A and C_B, and the mean compact composition is in the α solid solution (Tanzilli and Heckel, 1968). Also shown in Fig. 11 are measures by which the progress of homogenization can be characterized in two-phase systems:

(i) the diameter of the unstable β-phase spheres, ξ†;

(ii) the normalized solute concentration $C/\bar{C}$ at the position $r = L/2$, the minimum solute concentration in the system ($0 \leq (C/\bar{C})_{L/2} < 1$ if pure elemental powders are used).

Experimental verification of the two-phase mathematical model has been carried out on multilayer, finite diffusion couples fabricated by diffusion

---

†Although ξ will eventually decrease to zero if $\bar{C}$ is in the α-phase field, ξ may increase initially if large β-phase fluxes occur at the outset of the homogenization process (see Eq. (8)). This interface turnaround phenomenon has been studied analytically by Tanzilli and Heckel (1968) and experimentally by Heckel *et al.* (1972).

bonding alternate layers of foils of Ni and W ($\overline{C}$ in Ni-rich α solid solution) and Ag and Cu ($\overline{C}$ in both the Ag-rich and Cu-rich solid solutions) (Tanzilli and Heckel, 1971). This geometry was chosen to assure that the model and the experiment were identical in form. Available data for $\check{D}_\alpha$, $\check{D}_\beta$, $C_{\alpha\beta}$, and $C_{\beta\alpha}$ for both the Ni–W and Ag–Cu experiments were used for the computer calculations (see Tanzilli and Heckel, 1975, for detailed information on diffusivities; see Hansen, 1958, for solubilities). Good agreement between the mathematical models and experimental data for both alloy systems was achieved, thus validating the use of these modeling procedures in making quantitative predictions of diffusional kinetics in finite, multiphase situations.

It is clear from Fig. 11 that the values of ξ and $(C/\overline{C})_{L/2}$ as a function of homogenization time are determined by the interplay of a large number of parameters. Thus, calculations should be made for specific circumstances. However, the following parameters exert the primary influence on these measures of homogeneity:

for ξ: $\check{D}_\alpha$, l, t, geometry, and $C_{\alpha\beta}$;
for $(C/\overline{C})_{L/2}$: $\check{D}_\alpha$, l, t, geometry and $\overline{C}$.

It has been found empirically by Tanzilli and Heckel (1969) that values of (ξ/l) fall into a narrow band when plotted as a function of $(\check{D}_\alpha t/l^2)\cdot(C_{\alpha\beta}^Y)$, where $Y = 2.1$, 1.4, and 1.1 for planar, cylindrical, and spherical geometries, respectively. This correlation permits a reasonable approximation for ξ in terms of the major parameters which determine it (see Fig. 12). In general, scatter in the band is determined primarily by the ratio

$$\delta/\lambda = (C_{\beta\alpha} - C_{\alpha\beta})/[(C_B - C_{\beta\alpha})/(C_{\alpha\beta} - C_A)] \qquad (9)$$

as shown in Fig. 12. It has also been found empirically by Tanzilli and Heckel (1969) that a similar correlation exists for $(C/\overline{C})_{L/2}$ as shown in Fig. 12. (The nomograph portion of Fig. 12 permits graphical calculation of ξ/l and $(c/\overline{C})_{L/2}$ values in the manner indicated by the dotted lines.)

Figure 12 defines the approximate conditions for the loss of the β-phase:

$$(\check{D}_\alpha t/l^2)\, C_{\alpha\beta}^Y \cong 0.10 \qquad (\xi \to 0) \qquad (10)$$

and for the attainment of an essentially homogeneous α-phase

$$(\check{D}_\alpha t/l^2)\, \overline{C}^Y \cong 0.10 \qquad ((C/\overline{C})_{L/2} \to 1) \qquad (11)$$

The condition for Eq. (11) (i.e., $(C/\overline{C})_{L/2} \to 1$) is desired as an endpoint to the homogenization treatment, but is often difficult to determine experimentally. However, the time necessary to achieve the condition for Eq. (10) (i.e., $\xi \to 0$) may be readily determined experimentally and multiplied by

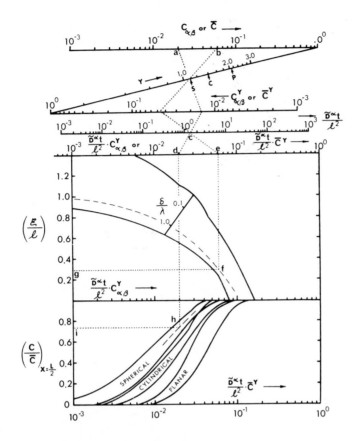

Fig. 12 Nomograph for approximating ξ/l and $(C/\overline{C})_{L/2}$ values as a function of time for two-phase homogenization. (Tanzilli and Heckel, 1969.)

$(\overline{C}/C_{\alpha\beta})^{\gamma}$ to obtain the time necessary to achieve essentially complete homogenization.

3. Three-Phase Model

The three-phase, concentric-sphere model can be defined in the manner used for the two-phase model with the addition of one more phase and one more interface (see Fig. 6 and Lanam and Heckel, 1971). The effects of the various diffusivity and solubility parameters were studied in detail by Lanam and Heckel (1971). The three-phase model was evaluated experimentally by Lanam and Heckel (1975), with multilayer, finite diffusion couples fabricated by diffusion bonding foils of Ni and Mo in the manner used for evaluation of the two-phase model. Good agreement between the numerical calculations and experimental data was achieved, thus validating the model and computational techniques.

The parameters which have the major effect on the loss of *both* of the second phases (β and γ in Fig. 6), i.e., the situation when $\zeta_\beta \to 0$, are the same as for the loss of the second phase in the two-phase model. The same situation holds for the major parameters which control the attainment of essentially complete homogenization, i.e., $(C/\overline{C})_{L/2} \to 1$. Figure 13 was thus formulated empirically for the three-phase model in the same manner as for the two-phase model, with ζ_2 indicating ζ_β.

4. Mechanical Working Effects

Mechanical working of the sphere–shell geometry was considered for extrusion or rolling by Heckel and Balasubramaniam (1971). It is assumed that both constituents will undergo plastic deformation during the mechanical working operation. Extrusion results in the change in the minor constituent sphere toward the shape of a cylinder; the cylinder radius decreases with increasing area reduction R due to extrusion. Rolling† results in the change in the minor constituent sphere toward the shape of a plate; the plate thickness decreases with increasing R. The shape change, when considered by itself, results in a decrease in the homogenization kinetics, as can be seen from the geometry parameters in the one-phase (Figs. 8 and 9; n parameter in Fig. 10), the two-phase (Y in Fig. 12), and the three-phase models (Y in Fig. 13). The reduction in particle dimensions with increasing R increases homogenization kinetics through the effect on l. The competition between geometry and dimension effects may be seen by reference to the one-phase model and Fig. 10. If both the sphere of minor constituent and the shell undergo the same deformation as the compact as a whole, an extrusion deformation of $R = 0.90$ (90% reduction in area) will transform the minor constituent spheres to prolate ellipsoids of revolution which closely approximate cylinders. Since the geometry parameter is changed from $n = 3$ to $n = 2$ (and the abscissa of Fig. 10 contains n^2), the geometry change considered alone requires an increase in processing time by a factor of 9/4 to achieve the same state of homogeneity. However, $R = 0.90$ would reduce the minimum particle dimension from l (sphere diameter) to $(1 - R)^{1/2}l$ (cylinder diameter). Since the particle dimension enters the abscissa of Fig. 10 as a squared term, an $R = 0.90$ reduction results in a thermal treatment time of only 10% of the spherical geometry thermal

---

†Whether by cross rolling or straight rolling is probably of little consequence since the idealized plate geometry (Fig. 5) is never achieved.

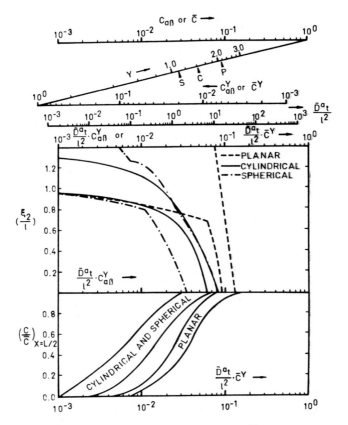

Fig. 13 Nomograph for approximating ξ_2/l (ξ_β/l) and $(C/\overline{C})_{L/2}$ values as a function of time for three-phase homogenization. (Lanam and Heckel, 1971.)

treatment time in order to achieve the same state of homogeneity. Thus, in total, the combined effects of geometry and dimension change give an overall homogenization time decrease to $(1/10) \cdot (9/4)$ or 9/40 the spherical model time.

The one-phase model (Fig. 10) has been adapted to mechanical working by Heckel and Balasubramaniam (1971) by adding the R parameter, as shown in Fig. 14. Geometry changes can be accounted for by appropriate changes in the n parameter of the ordinate scale. (Continuous or fractional changes in n could be used to accommodate the gradual change from one idealized geometry to another, rather than use the discontinuous change as indicated in the example above (i.e., from $n = 3$ to $n = 2$)). The dashed line in Fig. 14 indicates schematically the progress of homogenization in a series of alternating thermal and mechanical treatments. The first thermal treat-

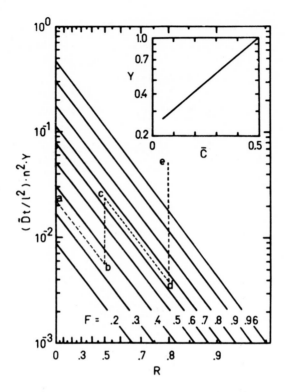

Fig. 14 Degree of interdiffusion F as a function of thermal and mechanical processing parameters for one-phase systems. Planar, $n = 1$; cylindrical, $n = 2$; spherical, $n = 3$. (Heckel and Balasubramaniam, 1971.)

ment ($R = 0$) results in $F = 0.35$ (point a) and is followed by a deformation treatment of $R = 0.50$. Since no interdiffusion occurs during working, the process follows the iso-F path to point b. Path b–c is a subsequent thermal treatment; c–d is another deformation, etc.

The two-phase model was also adapted to mechanical working by Heckel and Balasubramaniam (1971) using the approximations presented in Fig. 12. The effect of R on dissolution of the unstable α-phase is presented in Fig. 15; the effect of R on $(C/\bar{C})_{L/2}$ is presented in Fig. 16. The bands of uncertainty in Figs. 15 and 16 are carried over from Fig. 12 (i.e., $R = 0$) and are assumed to remain of constant width with increasing R. As was the case with Fig. 14, the R parameter is the fractional reduction in area of the dispersed minor constituent particles; equating R with the fractional reduction in area of the overall compact carries the implicit assumption that the matrix and dispersed particles undergo equal amounts of deformation.

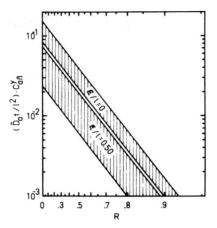

Fig. 15 Effect of thermal and mechanical processing parameters on the dissolution of β-phase during homogenization of two-phase systems. Planar, $y = 2.10$; cylindrical, $y = 1.40$; spherical, $y = 1.10$. (Heckel and Balasubramaniam, 1971.)

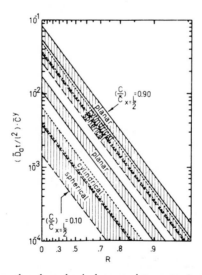

Fig. 16 Effect of thermal and mechanical processing parameters on the state of homogeneity of the α-phase matrix during the homogenization of two-phase systems. Planar, $y = 2.10$; cylindrical, $y = 1.40$; spherical, $y = 1.10$. (Heckel and Balasubramaniam, 1971.)

5. Mixing (Blending) and Particle Size Distribution Effects

The spherical (concentric-sphere) geometry (see Fig. 5) represents an ideal situation which is not attainable by blending and compaction due to the random nature of the mixing process. Likewise, the planar and cylindrical models are also ideal situations which cannot be attained by mechanical working of compacts. All of the mathematical models discussed above were defined for the case where all of the minor constituent particles in the compact have the same size l. Effects of the degree of mixing and of particle size variation have been explored analytically by Masteller *et al.* (1975); the one-phase, planar geometry system was used in order to simplify numerical calculations. The computational methods used were essentially the same as those employed for the development of the idealized one-phase model by Heckel and Balasubramaniam (1971).

Analytical mixing studies were carried out by considering solute plates, each of thickness l, separated by plates of solvent which could be specified to be of any thickness (i.e., nonquantized). The mean composition of the alternate-plate composite was in all instances $\overline{C} = 0.48$. Figure 17 gives examples of the uniform (idealized) distribution and a nonuniform distribution of the solute plates. Symmetry elements in the model contained two solute plates within the distance L; the positions $x = 0$ and $x = L$ were given the boundary conditions $\partial C/\partial x = 0$ to maintain $\overline{C} = 0.48$, for $0 \le x \le L$ for all time. Thus, the positions $x = 0$ and $x = L$ became mirror planes in the concentration–distance profiles, as shown in Fig. 17. Different states of mixing representing departures from the uniform (idealized) state were

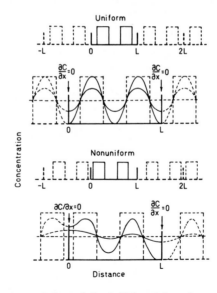

Fig. 17 Schematic representation of the initial and boundary conditions for uniform (condition A) and nonuniform particle (plate) distributions. (Masteller *et al.*, 1975.)

achieved by placing the two solute plates at various positions within the distance interval between $x = 0$ and $x = L$.

Figure 18 shows the effect on the concentration–distance profiles which result from departures from uniform mixing (case A). The time to achieve a given state of homogeneity increases with such departures. Two inter-related explanations for this are apparent in Fig. 18. First, solute concentrations at various locations in nonuniform distributions do not proceed directly to the mean ($\overline{C}$), but in fact can oscillate about the mean (e.g., case H, $x \cong 0.45\ L$; case M, $x = L$). This inefficiency in diffusional fluxes prolongs the homogenization process. Second, solute particles which are close together eventually begin to behave as if they were one large particle (i.e., the effective interdiffusion distance increases as the homogenization progresses). The distribution effects shown in Fig. 18 are also shown in Fig. 19. Obviously, the major effects of nonuniform mixing are to slow homogenization during the later stages of the process. The individual particles in the distributions considered in Figs. 18 and 19 initially behave independently as would uniformly distributed particles. More extreme departures from

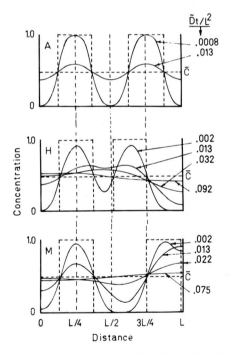

Fig. 18   Calculated concentration–distance profiles for uniform (condition A) and non-uniform (conditions H and M) initial conditions. $\overline{C} = 0.48$. (Masteller *et al.*, 1975.)

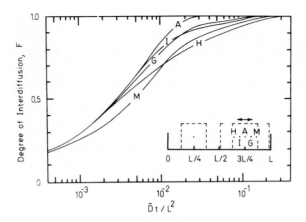

Fig. 19 Calculated degree of interdiffusion F for various initial particle (plate) distributions. $\overline{C} = 0.48$. (Masteller *et al.*, 1975.)

the uniformly dispersed case, of course, can lead to even larger reductions in homogenization kinetics (Masteller *et al.*, 1975).

Nonuniform particle sizes can also result in decreased homogenization kinetics. Figure 20 shows five different two-particle size distributions (all with the same mean size) and the effect of these distributions on the homogenization kinetics. In all instances, the kinetics are decreased, with the major effect occurring at the later stages of homogenization.

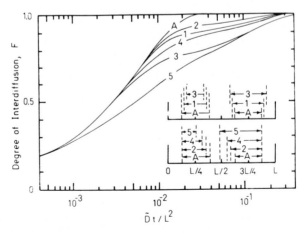

Fig. 20 Calculated degree of interdiffusion F for various particle size distributions all having the same mean particle size. $\overline{C} = 0.48$. (Masteller *et al.*, 1975.)

IV. EXPERIMENTAL EVALUATION OF MODELS

A. Experimental Techniques

Microstructural and compositional analysis techniques are necessary for complete evaluation of the predictive models discussed previously (Rudman, 1960). Quantitative metallography (Rhines and DeHoff, 1968) can be used to determine values of ξ for multiphase homogenization situations. For two-phase systems (Tanzilli and Heckel, 1975)

$$\xi/l = (V_v/V_{v_0})^{1/3} \tag{12}$$

where V_v and V_{v_0} are the volume fractions of β-phase after partial homogenization and initially, respectively. The volume fractions can be determined by point, line, or area counting (Heckel, 1975). For three-phase systems (Lanam *et al.*, 1975).

$$\xi_\gamma/l = (V_{v\gamma}/V_{v_0\gamma})^{1/3} \tag{13}$$

and

$$\xi_\beta/l = [(V_{v\gamma} + V_{v\beta})/V_{v_0\gamma}]^{1/3} \tag{14}$$

where ξ_γ and ξ_β are defined in Fig. 6, $V_{v\beta}$ and $V_{v\gamma}$ are the volume fractions of the β and γ phases after a given partial homogenization treatment, and $V_{v_0\gamma}$ is the initial volume fraction of the unstable phase. ξ_β is the same as ξ_2 in Fig. 13.

Compositional data on the mean concentration–distance profiles in par-

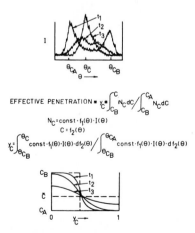

Fig. 21 Schematic representation of the treatment of X-ray line broadening peak profile data to yield composition as a function of effective penetration. (Masteller *et al.*, 1975.)

tially homogenized compacts may be obtained by X-ray compositional line broadening and electron microprobe techniques. (Energy dispersive X-ray analysis in conjunction with a scanning electron microscope can be used to provide a mapping of concentration contours.) X-ray compositional line broadening results from the spectrum of lattice parameters which exists in inhomogeneous solid solutions. Methods have been developed to define the frequency of occurrence of compositions in a solid solution and the mean effective concentration–distance profile (Heckel, 1964; Heckel and Balasubramaniam, 1971; Rudman, 1960). These techniques are summarized in Fig. 21 and provide the data necessary to evaluate experimentally the F, C_{min}, and C_{max} characteristics of one-phase systems (see Fig. 7) and the $(C/C)_{L/2}$ characteristics of multiphase systems (see Fig. 11). Concentration traces from electron microprobe data can also be analyzed to yield mean effective concentration–distance profiles as shown in Fig. 22.

B. Evaluation of Thermal Processing Models

1. *One-Phase System (Cu–Ni)*

The progress of homogenization in compacted blends of Cu and Ni powders can be readily observed from the change in X-ray diffraction peak profiles, since there is a variation in lattice parameter of about 0.08 Å across the width of the phase diagram (see Coles, 1956). A typical set of

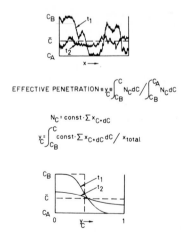

Fig. 22 Schematic representation of the treatment of electron microprobe data to yield composition as a function of effective penetration. (Heckel and Balasubramanian, 1971.)

Cu–Ni diffraction peaks as a function of thermal treatment is shown in Fig. 23 (Heckel *et al.*, 1970). The initial stage of homogenization is seen to be formation of the complete range of solid solutions from pure Cu to pure Ni; subsequently, the pattern sharpens into the single peak characteristic of the mean composition of the compact $\overline{C}$.

X-ray data of the type shown in Fig. 23 were used to define the F, C_{min}, and C_{max} parameters for a series of Cu–Ni compacted blends by Heckel and Balasubramaniam (1971). The experimental variables included $\overline{C}$, T, and l. Typical data for $\overline{C} = 0.20$ a.f. Cu (l being the Cu particle diameter) are compared to the one-phase concentric-sphere model predictions in Fig. 24 (D values were obtained from the literature (daSilva and Mehl, 1951) for the composition 0.20 a.f. Cu). Several features of Fig. 24 are worth noting:

(i)   the parameter $\tilde{D}t/l^2$ effectively normalizes the data to a band which narrows in width with increasing homogenization,

(ii)   the initial stages of homogenization proceed faster than predicted on the basis of volume diffusion data, indicating that diffusion via the extensive interconnected interparticle surfaces (and possible grain boundaries, as well) also contributes to mass transport,† and

(iii)   the later stages of homogenization proceed slower than predicted,

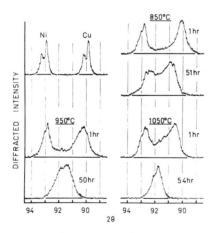

Fig. 23   X-ray diffraction peak profiles determined for Cu–Ni compacts given various thermal treatments. The (311) peaks as determined using CuK$_\alpha$ radiation are shown for mean compact positions of $\overline{C} = 0.52$ a.f. Ni. Cu 100/140 mesh. Ni 100/140 mesh. (Heckel and Balasubramanian, 1971.)

† The fact that fine matrix (Ni) particle sizes and low temperatures give the fastest initial homogenization rate provides additional support for the surface diffusion hypothesis.

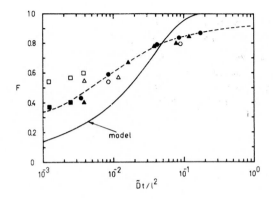

Fig. 24 Comparison of experimental F data for compacted blends of Cu and Ni powders ($\overline{C} = 0.20$ a.f. Cu) to concentric-sphere model predictions. □ ■ 850°C, △ ▲ 950°C, ○ ● 1050°C. Cu 200/270 ($l = 65$ μm). Closed points—coarse Ni, open points—fine Ni. (Heckel and Balasubramanian, 1971.)

probably due to the effects of imperfect mixing and/or small but significant ranges of Cu particle sizes about the mean (l).

Figure 25 compares the range of Cu concentrations (C_{min} and C_{max}) as determined by the concentric-sphere model and by experimental studies on the same specimens used in Fig. 24. As in Fig. 24, the parameter Dt/l^2 normalizes the data into narrow bands. In addition, progress of the later stages of

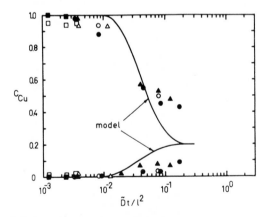

Fig. 25 Comparison of experimental composition range data (C_{min} and C_{max}) for compacted blends of Cu and Ni powders ($\overline{C} = 0.20$ a.f. Cu) to concentric-sphere model predictions, □ ■ 850°C, △ ▲ 950°C, ○ ● 1050°C. Cu 200/270 ($l = 65$ μm). Closed points—coarse Ni, open points—fine Ni. (Heckel and Balasubramanian, 1971.)

homogenization is slower than predicted, as was the case in Fig. 24. Also, initial homogenization of the Cu-rich regions (C_{max} curve) is faster than predicted. Although this is probably the result of surface diffusion (as discussed above), the apparent magnitude of the effect may be accentuated somewhat by the restriction of the model to a concentration-independent interdiffusion coefficient. Considerable concentration dependence of $\tilde{D}$ exists in the Cu–Ni system, the values for Cu-rich alloys being the largest. Thus, the Cu-rich regions would tend to homogenize more rapidly than would be predicted with a $\tilde{D}$ value characteristic of the mean composition. (The higher rate of homogenization of Cu-rich regions of the compacts is also apparent in the X-ray peak profiles shown in Fig. 23; the Cu-rich peaks broaden more rapidly than the Ni-rich peaks.)

A wide range of homogenization data are presented for the Cu–Ni system in Fig. 26, which includes variations in $\overline{C}$ and l as well as the parameters included in Fig. 24. The l values in Fig. 26 are for Cu particle sizes in Ni-rich compacts and vice versa. It is significant that, even though the data band appears to be large, a plot of these same data in the form of F versus t would range over two and one-half orders of magnitude in t in the range of $0.6 < F < 0.8$. Thus, the abscissa parameter is effective in describing the major parameters in the one-phase model. Nevertheless, factors such as

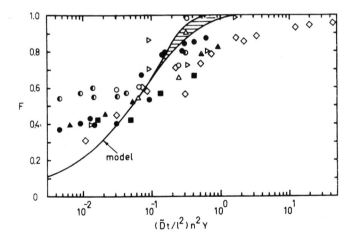

Fig. 26 Comparison of a wide range of Cu–Ni homogenization data to the normalized concentric-sphere model (Fig. 10). Powder compacts for $n = 3$: ● ◐ Ni-0.20 a.f. Cu, ▲ Ni-0.30 a.f. Cu, ■ Ni-0.40 a.f. Cu ($l - $ Cu); ▷ Cu-0.11 a.f. Ni. ○ Cu-0.21 a.f. Ni. △ Cu-0.32 a.f. ni, ◇ Cu-0.52 a.f. Ni ($l - $ Ni). 47 μm $\leq l \leq$ 122 μm. 850°C $\leq T \leq$ 1050°C. (Heckel and Balasubramanian, 1971.)

surface diffusion, blending, etc., can play a significant role in the overall kinetics of homogenization.

2. Two-Phase System (Ni–W)

The progress of homogenization in compacted blends of Ni and W powders can be studied by quantitative metallography to obtain data on dissolution of the β-phase (i.e., application of Eq. (12) to photomicrographs like those shown in Fig. 2). (Special quantitative metallography techniques (DeHoff, 1962) can be used for situations where data on the behavior of a

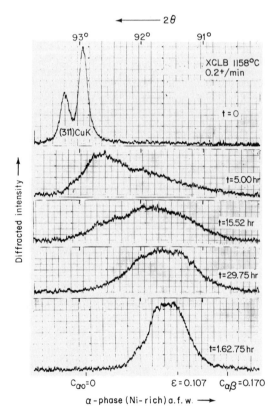

Fig. 27 X-ray diffraction peak profiles of the α-phase of Ni–W compacted blends given various thermal treatments. The (311) peaks as determined using CuK_α radiation are shown for mean compact compositions of $\overline{C} = 0.107$ a.f. W. A composition scale extending from zero to $C_{\alpha\beta}$ (0.17 a.f. W) is shown for convenience in interpreting the 2θ diffraction scale. Ni −400 mesh, W 325/400 mesh. (Tanzilli and Heckel, 1975.)

particle size distribution are desired.) In addition, X-ray compositional line broadening can be used even though two solid solutions of limited extent exist in the microstructure. Figure 27 indicates the change in shape and position of the (311) diffraction peak of the Ni-rich α solid solution as the W-rich particles dissolve during homogenization. Initially, the peak spreads over the entire composition range of the solid solution; later, the peak sharpens at the mean composition of the compact. The low-W end of the peak is indicative of the value of $C_{L/2}$ necessary to evaluate the two-phase concentric-sphere model.

Homogenization data in terms of ξ/l and $(C/\overline{C})_{L/2}$ for Ni–W compacts with $\overline{C} = 0.17$ a.f. W are presented in Figs. 28 and 29. Available data on solubilities and diffusion coefficients (see Tanzilli and Heckel, 1971, 1975) were used for the model calculations and data plotting. The ξ/l data fit the predicted curve fairly well except for the low-temperature data points; it is suspected that interdiffusion via interparticle surfaces in the matrix serves to enhance the overall homogenization kinetics for these specimens. The

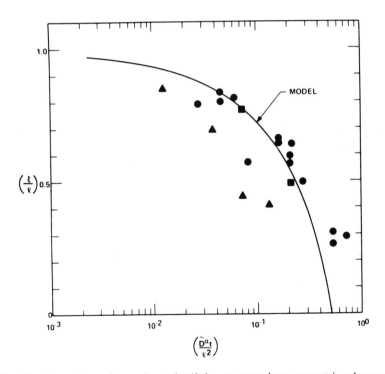

Fig. 28 Comparison of experimental ξ/l data to two-phase concentric-sphere model predictions for Ni–W compacted blends having $\overline{C} = 0.107$ a.f. W. ▲ 1156°C, ● 1207°C, ■ 1271°C. (Tanzilli and Heckel, 1969.)

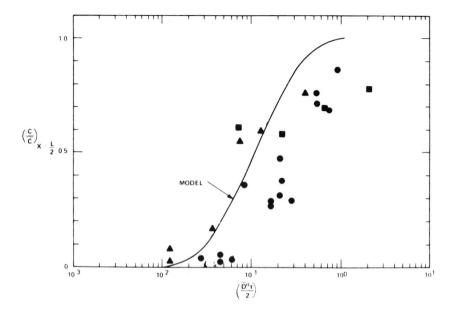

Fig. 29 Comparison of experimental $(C/\overline{C})_{L/2}$ data to two-phase concentric-sphere model predictions for Ni–W compacted blends having $\overline{C} = 0.107$ a.f. W. ▲ 1156°C, ● 1207°C, ■ 1271°C. (Tanzilli and Heckel, 1969.)

$(C/\overline{C})_{L/2}$ data indicate slower homogenization than that predicted; this is presumably due to the effects of nonuniform mixing. The diffusional geometry exhibited in Fig. 3, although generally spherical in form, shows numerous examples of the effects of nonuniform mixing in the homogenization of Ni–W compacts.

3. Three-Phase System (Ni–Mo)

The progress of homogenization in compacted blends of Ni and Mo powders can be observed by the same techniques used for the two-phase Ni–W system, except that consideration must be given to the presence of the MoNi intermediate phase. Figure 30 shows the interrelation between the Ni–Mo phase diagram, the concentration–distance profile, and the concentric-sphere model; the initial stage of homogenization (three phases) is shown (see Fig. 4 for the corresponding photomicrograph) for the situation where $\overline{C}$ is in the Ni-rich α solid solution field. Equations (13) and (14) can be used to obtain measures of the ξ_γ and ξ_β which are depicted in Fig. 30 as ξ_1 and ξ_2, respectively.

The diameters of the Mo-rich γ-phase (ξ_1) and the β–γ composites (ξ_2)

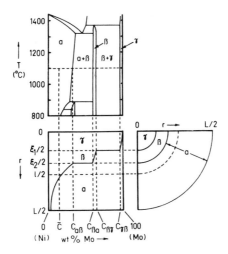

Fig. 30 Schematic representation of the three-phase concentric-sphere model for the Ni–Mo system. (Lanam *et al.*, 1975.)

were determined experimentally for a wide range of temperatures, initial Mo particle sizes (l), and mean compact compositions ($\overline{C}$). These data are compared to the concentric-sphere model predictions in Figs. 31 and 32. (Literature data were used for the interdiffusion coefficients and solubilities

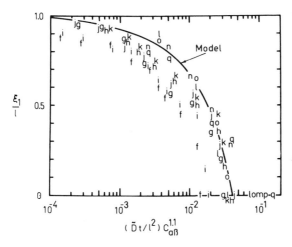

Fig. 31 Comparison of experimental ξ_1/l (ξ_γ/l) data to three-phase concentric-sphere model predictions for Ni–Mo compacted blends. 17.2 μm $\leq l \leq$ 48 μm, 950°C $\leq T \leq$ 1200°C, 5 wt% Mo $\leq$ Mo $\leq$ 15 wt% Mo, 60 sec $\leq t \leq$ 180,000 sec. (Lanam *et al.*, 1975.)

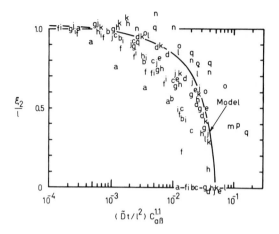

Fig. 32 Comparison of experimental ξ_2/l (ξ_β/l) data to three-phase concentric-sphere model predictions for Ni–Mo compacted blends. 8.4 μm $\leq l \leq$ 48 μm, 950°C $\leq T \leq$ 1250°C 5 wt% Mo $\leq \bar{C} \leq$ 15 wt% Mo, 30 sec $\leq t \leq$ 180,000 sec. (Lanam *et al.*, 1975.)

and are presented by Lanam *et al.* (1975)). The ξ data agree closely with the model prediction except for the low-temperature (950°C) points (f and i) which show the typical enhanced homogenization kinetics. The ξ_2 data exhibit somewhat more scatter about the model prediction line; the low-temperature points (a, f, and i) show enhanced kinetics, and the high $\bar{C}$ points (o, p, and g) generally show retarded homogenization kinetics. The latter effect is probably the result of a blending problem which leads to local regions in the compact where the composition of solute exceeds the α-phase solubility, thus requiring longer times for complete dissolution of the MoNi β-phase. This effect is not found for the dissolution of the Mo-rich γ-phase particles since the β-phase effectively screens them from the effects of adjacent γ-phase particles that are not dispersed properly. Larger mean compact compositions of $\bar{C} = 0.30$ a.f Mo, which lie in the Ni-rich α solid solution, have been shown to exhibit very slow β-phase dissolution during homogenization due to this blending effect, even though the γ-phase dissolution proceeded in accord with the model predictions.

4. Mechanical Working

The mechanical working model discussed previously for one-phase systems (Heckel and Balasubramaniam, 1971) was evaluated by studies of compacts of compacts of Ni and Cu powders ($\bar{C} = 0.20$ a.f. Cu) which were given alternate thermal and cold rolling treatments. Degree of interdiffusion data (F) for comparison to model predictions (Fig. 14) were obtained by X-ray compositional line broadening analysis following the thermal treatments.

The development of homogeneity through the alternate thermal and rolling treatments is shown in Figs. 33 and 34 which have F values of 0.65 and 0.76, respectively. The etching response provides a relatively sensitive qualitative indication of the state of homogeneity in these materials.

A comparison between the mechanical working model and experimentally determined F values is presented in Fig. 35. The reduction in size of the disperesed Cu particles by rolling was assumed to be identical to that of the

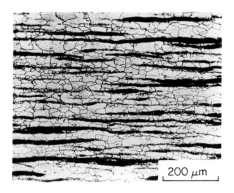

Fig. 33 Microstructure of a compacted blend of Ni and Cu powders ($\overline{C}$ = 0.20 a.f. Cu) given a sequence of thermal and rolling (total R = 0.81) treatments to yield an F value of 0.65 (specimen 41 in Table I and Fig. 35). (Heckel and Balasubramanian, 1971.)

Fig. 34 Microstructure of a compacted blend of Ni and Cu powders ($\overline{C}$ = 0.20 a.f. Cu) given a sequence of thermal and rolling (total R = 0.95) treatments to yield an F value of 0.76 (specimen 42 in Table I and Fig. 35). (Heckel and Balasubramanian, 1971.)

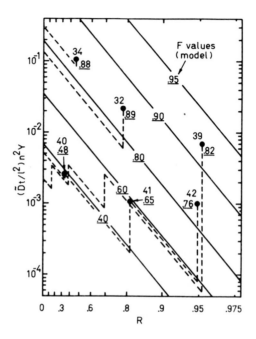

Fig. 35 Summary of F data obtained from X-ray compositional line broadening measurements on compacted blends of Ni and Cu powders ($\overline{C} = 0.20$ a.f. Cu) given four different sequences of thermal and rolling treatments. $F_{R=0}$ from experiment; $n = 1$ for $R > 0$. Photomicrographs of specimens 41 and 42 are shown in Fig. 33 and 34. (Heckel and Balasubramanian, 1971.)

overall compacts. Thus, R values were taken to be the total reduction in thickness of the compact. The dashed traces indicate the four separate thermal/mechanical paths studied; individual points are marked with the experimental F values (underlined numbers) and the specimen numbers. The origins of each of the paths at $R = 0$ were taken to be the experimental values of F determined after the initial thermal treatment rather than the values predicted by the model. This procedure eliminated the influence of any discrepancies between the concentric-sphere thermal model and experimental data on the evaluation of the mechanical working model. In addition, the value of the geometry parameter n was taken to be unity for all working treatments rather than allowing it to change gradually from $n = 3$ (spherical) to $n = 1$ (planar) during the progression of cold rolling treatments.

An indication of the accuracy of the mechanical working model can be obtained by comparison of the experimental F data and the model values taken from the iso-F diagonal lines shown in Table I. It may be seen that the experimental data are all within the ± 0.10 of the predicted F values and

TABLE I

Comparison of the Experimental and Model-Predicted F *Values for the Four Thermal/Mechanical Processing Paths Shown in Fig. 35* [a]

Path	Data point number	F values	
		Experimental	Predicted
1	39	0.82	0.92
2	40	0.48	0.38
	41	0.65	0.59
	42	0.76	0.83
3	32	0.89	0.84
4	34	0.88	0.83

[a]The paths are listed in order of lowest to highest F value at the position $R = 0$ in Fig. 35.

most of the values are within ± 0.07. Thus, the mechanical working model appears to describe the interactions of the various parameters involved and the overall predictive accuracy is equivalent to that found for the one-phase concentric-sphere thermal model.

V. HOMOGENIZATION PROCESS DESIGN

The foregoing comparisons between predictive models and experimental data for homogenization indicate that the models can be used to provide guidelines for process design even though factors such as improper blending, variations in minor constituent particle size, and contributions due to surface diffusion (which are not treated by the models) can lead to departures from predicted homogenization rates. In addition, the models can be used to understand the interplay between the various parameters considered by the models and to define experiments which would empirically improve the predictive capabilities of the models. For example, an appropriate model could be used to analyze a set of experimental homogenization data in order to obtain an effective interdiffusion coefficient which would contain the contributions of both volume (lattice) and surface transport.

In general, it has been found that the mathematical models provide rather precise predictions of complex phenomena for the situations where the model and experimental geometries are exactly the same and the nature of the transport process is by volume diffusion (Tanzilli and Heckel, 1971; Lanam and Heckel, 1975; Hickl and Heckel, 1975); inaccuracies of various magnitudes develop when this is not the case. However, as is indicated above, it is possible to define effective parameters for the models in order to provide agreement for special circumstances.

The three-phase model of Lanam and Heckel (1971) was used to provide processing guidelines for the laboratory scale powder fabrication of Hastelloy B sheet by Heckel and Erich (1972). This study included Ni-20 wt% Mo and Ni-20 wt% Mo-4 wt% Fe alloy compositions as well as the nominal Hastelloy B composition (Ni-28 wt% Mo-5 wt% Fe). Commercially available powders of Ni, Mo, and Fe were used, and the processing conditions included blending, compaction, and alternate cycles of thermal and rolling treatments ending with a thermal treatment in order to obtain an annealed condition.

The processing conditions were defined by considering the binary Ni–Mo and Ni–Fe systems since it was anticipated that both the Mo and Fe powders would be surrounded by Ni in the Ni–Mo–Fe alloys. Ternary interactions were not taken into account. Essentially complete homogenization for the three-phase Ni–Mo system (Fig. 13) in terms of $(C/\overline{C})_{L/2}$ approaching unity may be defined from

$$(\tilde{D}_\alpha t / l^2)\overline{C}^{1.1} \cong 0.08 \qquad (15)$$

Similarly, for the one-phase Ni–Fe system (Fig. 10), essentially complete

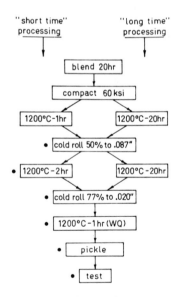

Fig. 36 Summary of the processing routes for the powder fabrication of Hastelloy B by homogenization of compacted blends of powders.

homogeneity (in terms of F approaching unity) may be defined from

$$(\check{D}t/l^2)n^2 Y \cong 2.0 \qquad (16)$$

where $n = 3$ and $Y \cong 0.25$ (for $\overline{C} \cong 0.05$). Scanning electron microscopy showed that the Mo and Fe powder particles were less than 5 μm in diameter; this value was used for l. The furnace temperature chosen was 1200°C (2200°F) since this is the maximum which would be readily attained with standard equipment. Interdiffusion coefficients for the Ni–Mo and Ni–Fe systems (Lanam *et al.*, 1975; Lanam and Heckel, 1975; Goldstein *et al.*,

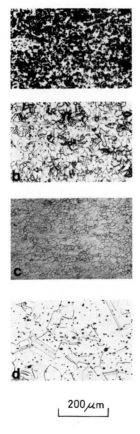

200 μm

Fig. 37 Microstructures during various stages of the short-time powder processing route given in Fig. 36. The cast and wrought alloy microstructure is shown for comparison. (a) Compacted 60 ksi + 1200°C – 1 hr. (b) Stage (a) + CR 50% + 1200°C – 2hr. (c) Stage (b) + CR 77% (89% total) + 1200°C – 1hr (WQ) (S28–5). (d) CW Hastelloy B + CR 77% + 1200°C – 1hr (WQ) (CCW1).

1965; Darwin *et al.*, 1963) at 1200°C (2200°F) along with l and $\overline{C}$ values, when substituted into Eqs. (15) and (16), indicated that homogenization would be complete in each of the binary systems after about 15 to 30 min at temperature. Allowance was made for inaccuracies in input parameters, improper mixing, etc., by extending the total time to 4 hr. The acceleration of homogenization by cold working was not considered in the time calculation and was viewed as an additional safety factor.

The overall processing route for the 1200°C-4 hr thermal treatment is shown in Fig. 36 under the heading "short time" (coded S in subsequent figures). Two control routes are also shown in Fig. 36:

(i)   a long time powder fabrication route (coded L in subsequent figures) in which the total thermal treatment was 1200°C-41 hr, and

(ii)   a finishing treatment for commercial cast and wrought (CCW) Hastelloy B.

Figure 37 shows the development of the microstructure in the short-time processed Ni-28 Mo-5 Fe material (S 28–5) and a comparison to the CCW

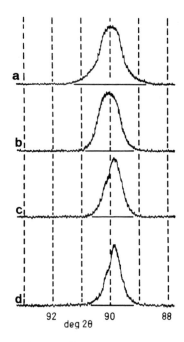

Fig. 38   The development of compositional homogeneity in the short-time powder processing route as indicated by the narrowing of the (311) X-ray diffraction peak of the alloy. The cast and wrought alloy peak is shown for comparison. (a)–(d) Same as in Fig. 37.

Hastelloy B product. Grain size in the powder fabricated alloy is considerably finer than that in the cast and wrought alloy. Sharpening of the X-ray diffraction profile of the (311) peak for the microstructures shown in Fig. 37 may be seen in Fig. 38; the peak of the powder fabricated product is essentially the same as that of the cast and wrought product.

Figures 39 and 40 show the mechanical properties and corrosion resistance of the powder fabricated alloys. The short-time processed Hastelloy B (S 28–5) can be seen to have slightly higher strength than the cast and

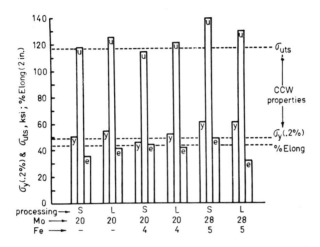

Fig. 39 Mechanical properties of alloys produced by homogenization of compacted blends of powders. Cast and wrought (CCW) properties are shown for comparison.

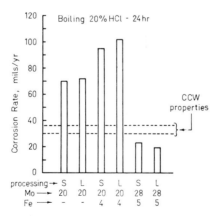

Fig. 40 Corrosion rate of alloys produced by homogenization of compacted blends of powders. Cast and wrought (CCW) alloy corrosion rates are shown for comparison.

wrought Hastelloy B, probably due to the finer grain size; the ductility is the same. The corrosion resistance of the powder fabricated alloy is better than that of the cast and wrought material, presumably due to the lower concentration of tramp impurity elements.

VI. SUMMARY AND CONCLUSIONS

The use of homogenization processing for the fabrication of alloys by powder metallurgy has many attractive advantages. However, if the potential of this type of processing is to be exploited, detailed consideration must be given to the design of processes which assure adequate homogeneity in the powder metallurgy product. Such process design must include proper specification of thermal and mechanical working treatments to provide the homogeneity required in the final product, optimization of processing parameters to define the most economical processing route, and characterization of the state of homogeneity in the alloy and at various critical points along the processing route to assure that the process design was effective.

Homogenization processing kinetics are determined by a large number of parameters which have complex interactions. The mathematical models presented in this chapter provide an insight to the effects of these parameters and their interactions. Furthermore, the models can provide predictive guidelines for the design of processes; comparisons between the model predictions and experimental data will serve to define the probable accuracy of such predictions. The experimental techniques discussed can provide a means for evaluating the effectiveness of the models in specific processing situations.

REFERENCES

Baty, D. L., Tanzilli, R. A., and Heckel, R. W. (1970). *Met. Trans.* **1**, 1651.
Carslaw, H. S., and Jaeger, J. C. (1959). "Conduction of Heat in Solids." Oxford Univ. Press, London and New York.
Chevenard, P., and Wache, X, (1944). *Rev. Met. (Paris)* **41**, 353.
Coles, B. R. (1956). *J. Inst. Met.* **84**, 346.
Darwin, A., Leroy, V., Contsouradis, D., and Habraken, L. (1963). *Rev. Met. (Paris)* **60**, 275.
daSilva, L. C. C., and Mehl, R. F. (1951). *Trans. AIME* **191**, 155.
DeHoff, R. T. (1962). *Trans. Met. Soc. AIME* **224**, 474.
Duwez, P., and Jordan, C. B. (1949), *Trans. ASM* **41**, 194.
Fisher, B., and Rudman, P. S. (1961). *J. App. Phys.* **32**, 1604.
Gertziken, S. D., and Feingold, M. (1940). *Zh. Theor. Fiz.* **10**, 574.
Goldstein, J. I., Hanneman, R. E., and Ogilvie, R. E. (1965). *Trans. Met. Soc. AIME* **233**, 812.

Hansen, M. (1958). "Constitution of Binary Alloys." McGraw-Hill, New York.

Heckel, R. W. (1964). *Trans. ASM* **57**, 443.

Heckel, R. W. (1975). *In* "Microstructural Science" (P. M. French, R. J. Gray, and J. H. McCall, eds.) Vol. 3, p. 671. Elsevier, New York.

Heckel, R. W., and Balasubramaniam, M. (1971). *Met. Trans* **2**, 379.

Heckel, R. W., and Erich, D. L. (1972). *In* "1971 Fall P/M Conference Proceedings" (S. Mocarski, ed.), p. 281. Metal Powder Industries Federation, Princeton, New Jersey.

Heckel, R. W., Lanam, R. D., and Tanzilli, R. A. (1970). *In* "Perspectives in Powder Metallurgy, Vol. 5" (J. S. Hirschhorn and K. H. Roll, eds.), p. 139. Plenum Press, New York.

Heckel, R. W., Hickl, A. J., Zaehring, R. J., and Tanzilli, R. A., (1972). *Met. Trans* **3**, 2565.

Hickl, A. J., and Heckel, R. W. (1975). *Met. Trans.* **6A**, 431.

Koczak, M. J. (1976). *Int. J. Powder Met. Powder Tech.* **12**, 133.

Lanam, R. D., and Heckel, R. W. (1971). *Met. Trans.* **2**, 2255.

Lanam, R. D., and Heckel, R. W. (1975). *Met. Trans.* **6A**, 421.

Lanam, R. D., Yeh, F. C. H., Rovsek, J. E., Smith, D. W., and Heckel, R. W. (1975). *Met. Trans.* **6A**, 337.

Lund, J. A., Krantz, T., and Mackiw, V. N. (1960). *In* "Progress in Powder Metallurgy" p. 160. Metal Powder Industries Federation, Princeton, New Jersey.

Masteller, M. S., Heckel, R. W., and Sekerka, R. F. (1975). *Met. Trans.* **6A**, 869.

Murray, D., and Landis, F. (1959). *Trans. ASME* **81D**, 106.

Purdy, G. R., and Kirkaldy, J. S. (1971). *Met. Trans.* **2**, 371.

Raichenko, A. I. (1961). *Fiz. Metal. Metalloved.* **11**, 49.

Raichenko, A. I., and Fedorchenko, I. M. (1958). *Dokl. Akad. Nauk. Ukr. SSR* **3**, 255.

Rhines, F. N., and Colton, R. A. (1942). *Trans. ASM* **30**, 166.

Rhines, F. N. and DeHoff, R. T. (1968). "Quantitative Metallography." McGraw-Hill, New York.

Rudman, P. S. (1960). *Acta Cryst.* **13**, 905.

Sekerka, R. F., Jeanfils, C. L., and Heckel, R. W. (1975). *In* "Lectures on the Theory of Phase Transformations" (H. I. Aaronson, ed.), p. 117. Met. Soc. AIME, New York.

Smigelskas, A. D., and Kirkendall, E. O. (1947). *Trans. AIME* **171**, 130.

Tanzilli, R. A., and Heckel, R. W. (1968). *Trans. Met. Soc. AIME* **242**, 2313.

Tanzilli, R. A., and Heckel, R. W. (1969). *Trans. Met. Soc. AIME* **245**, 1363.

Tanzilli, R. A., and Heckel, R. W. (1971). *Met. Trans.* **2**, 1779.

Tanzilli, R. A., and Heckel, R. W. (1975). *Met. Trans.* **6A**, 329.

Wagner, C. (1960). *In* "Diffusion in Gases, Liquids and Solids" (W. Jost), p. 68. Academic Press, New York.

Wienbaum, S. (1948). *J. App. Phys.* **19**, 897.

Chapter 4

Deformation Processing of Sintered Powder Materials

Howard A. Kuhn

DEPARTMENT OF METALLURGICAL AND MATERIALS ENGINEERING
UNIVERSITY OF PITTSBURGH
PITTSBURGH, PENNSYLVANIA

I. INTRODUCTION

Production of parts by conventional powder metallurgy methods involves compaction and sintering. The resulting material has a substantial volume fraction of voids (interstices between powder particles) which limit its use to less than heavy duty applications. One method to enhance the properties of sintered powder materials involves deformation processing, which simultaneously densifies the material and develops the final desired part shape.

Fabrication of structural parts and components by deformation processing of sintered powder materials is a process of considerable importance in modern manufacturing technology. Forging of powder preforms is presently a production reality for high-strength machine parts, extrusion is used for production of superalloy barstock, and rolling of compacted powder is undergoing pilot line study. Powder forging is particularly attractive because it blends the cost and material saving advantages of conventional, press-and-sinter powder metallurgy with the high production rates and property enhancement of forging.

Powder forging involves fabrication of a preform by the usual press-and-sinter process, followed by forging of the porous preform into the final shape. Forging is generally performed in one blow in confined dies to eliminate flash formation and achieve net shapes. Thus, parts containing through holes and complex configurations may be manufactured with little or no material loss due to trimming, hole punching, machining, or grinding.

Plastic deformation of sintered powder materials is similar to that of conventional, fully dense materials, but there are additional complications due to the substantial volume fraction of voids (interstices between powder particles) in the preform material. In particular, the voids must be eliminated during deformation so that a sound metallurgical structure is obtained. The voids, however, are sites of weakness at which ductile fractures may initiate during deformation. Finally, the existence of voids leads to volume change, so that die design practice cannot be approached through conventional guidelines, and classical plasticity theory cannot be used for plasticity analysis.

In this chapter, a rational approach is developed for determination of the process parameters for successful forging of powder preforms. Experimental and theoretical analysis of the fundamental behavior of sintered powder material is first presented and then followed by examples of application.

II. PLASTIC DEFORMATION OF SINTERED POWDER METAL

The fundamental mechanical responses of porous metal, namely densification, plastic flow, and fracture, are best determined through compression tests on cylindrical specimens. Proper manipulation of the test conditions permits development of a wide range of stress and strain states typical of those in actual forming processes.

A. Densification

1. *Physical Model*

Investigation of densification of a porous metal is facilitated by consideration of deformation of a material element containing a void. It is well known from plasticity analysis of a thick-walled sphere (Torre, 1948) that it is impossible to completely close a hole by hydrostatic pressure of finite magnitude (illustrated in Fig. 1). The pressure required for plastic deformation of a sphere containing a hole is given by

$$p = 2\sigma_0 \ln r_0/r_i \qquad (1)$$

where σ_0 is the flow stress of the material, r_0 the outside radius (equivalent to mean space between voids), and r_i is the hole radius (equivalent to void radius). It is clear that voids of large diameter (large r_i) require less pressure for densification than small voids, and that, as the void diameter approaches zero, the pressure required for densification becomes unbounded. Under hydrostatic pressure, the void simply changes size, but not shape, since the pressure is equal in all directions.

If the stress state imposed on the void consists of a deviatoric component as well as a hydrostatic pressure component, the void will become flattened somewhat, and collapse to zero volume is possible (Fig. 1(b)). A deviatoric component of the stress state results when there are differences in principal stresses in the three coordinate directions, which lead to shear and the resulting shape change of the void. For large values of deviatoric stress (Fig. 1(c)) the void becomes extremely flattened and elongated. Under these conditions, the void may first be separated into smaller voids. Also, opposite sides of the collapsed void may shear along their interface.

The pure hydrostatic stress state (Fig. 1(a)) is obviously not conducive to

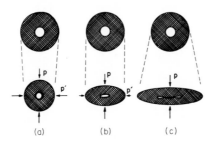

Fig. 1 Illustration of void deformation under (a) hydrostatic pressure, $p' = p$ (isostatic pressing); (b) repressing, $p' < p$ (no lateral deformation); (c) true forging, $p' = 0$ (lateral flow).

complete densification of a porous metal. This type of stress state is achieved in an isostatic pressure chamber, as for isostatic compaction of loose powder (Fig. 2(a)).

The stress state shown in Fig. 1(b) consists of a small difference between vertical and horizontal stresses and results in very little movement in the horizontal direction (small lateral flow). This type of stress state exists in the repressing type of deformation (Fig. 2(a)), where material is constrained from moving in the lateral direction by die walls. Some densification may be achieved, but as full density is approached, the stress state in this case approaches a pure hydrostatic condition, as in Fig. 1(a).

The stress state shown in Fig. 1(c) occurs when the material expands (flows laterally) freely during compression, as in upsetting (Fig. 2(c)). This type of stress state and deformation is obviously very beneficial for complete densification. Note that, at the later stages of upsetting, the material reaches

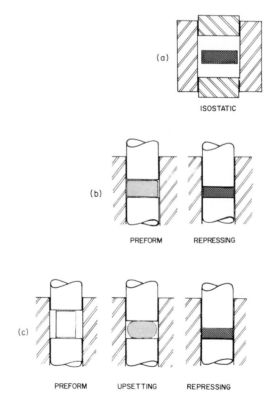

Fig. 2 Overall deformation in (a) isostatic pressing; (b) repressing; (c) forging (upsetting in a trap die).

the die walls, and further compression occurs in a repressing type of deformation (Fig. 2(b)). This, in turn, leads to hydrostatic pressure (Fig. 2(a)) as full density is approached. Because of the difficulties in achieving full density in the isostatic or repressing modes of deformation, it is clear that it is most beneficial to develop full density in an upsetting (free flowing) type of deformation before the repressing and hydrostatic stress states are developed in the later stages of the deformation process.

The model of void behavior given in this section is a pure mechanical model, devoid of any considerations of densification by diffusion transport of metal. Although the latter mechanisms are active in long-time processes such as hot isostatic pressing and hot pressing, where strain rates are 10^{-4} to 10^{-1} sec^{-1}, they have a negligible effect in conventional forging, where strain rates of 10^1 to 10^2 sec^{-1} are typical. In this chapter, we will be concerned only with conventional forging rates, and the mechanical model of densification will suffice. It should be pointed out that investigation of void closure by combined mechanical and diffusion-related deformation in the intermediate strain rate range, 10^{-1} to 10^1 sec^{-1}, is an open area for research.

2. Densification and Properties

The role of pores in limiting mechanical properties of materials is obvious. Voids act as sites for initiation of fractures and provide an easy path for crack propagation. Elimination of pores in the preform by deformation processing is imperative for achievement of high-performance properties.

This subject is treated in detail in Chapter 5. It is shown that static mechanical properties (e.g., tensile strength and ductility) are directly related to residual porosity. The dynamic mechanical properties (fatigue and impact resistance), however, are affected not only by amount of residual porosity but, at full density, by the characteristics of metal flow during the deformation process. Specifically, reaching full density by means of extensive lateral flow results in greater properties than by means of repressing deformation. Lateral flow involves shearing, as depicted in Figs. 1(c) and 2(c), while repressing involves pore collapse, as depicted in Figs. 1(b) and 2(b).

Details and discussion of these results are given in Chapter 5. The relevance of these results in deformation process design is given in later sections of the present chapter.

3. Uniformity of Densification

Densification, or closing up of voids by plastic deformation, is strongly affected by the stress state in the material, as shown in Section II.A.1. During most deformation processes, however, the stress state is not uniform throughout the workpiece, and densification, consequently, is not uniform. In simple compression of a cylinder, for example (illustrated schematically in Fig. 3), friction at the die contact surfaces causes a radial compressive stress which, combined with the axial compressive stress from the die, leads to a nearly pure hydrostatic stress state near the surface. This results in a cone of relatively undeformed material, commonly known as a dead-metal zone, at each die contact surface. Bounding these cones of undeformed material are bands of material deformed under very high shear stresses. These bands traverse the sample diagonally from corner to corner and form a shear cross. The original cylindrical surface, meanwhile, expands non-uniformly, leading to a barreled surface. Stresses at the barreled surface consist of compression in the axial direction and tension in the circumferential direction. Reducing friction reduces the nonuniformity of deformation. Further discussion of this aspect of the compression test will be given in Section II.C.

The effects of nonuniform stress and deformation on densification, indicated in the micrographs of Fig. 4(a), are taken from a cylinder upset with no lubrication. In the dead-metal zones, little deformation has occurred and the voids remain relatively undeformed and spherical due to the nearly pure hydrostatic stress state (see Fig. 1(a)). In the shear zones, however, voids are considerably flattened and the increase in density is great (see Fig. 1(c)). In the barreled outer regions, tensile stresses have enlarged the voids. Micrographs in Fig. 4(b), taken from a cylinder upset with good lubrication, show much more uniform deformation and densification. Throughout the sample, voids are flattened and elongated. It is again evident, that a large deviatoric stress component (shear stress) is beneficial in enhancing densification, as shown in Fig. 4(b) and the shear region of Fig. 4(a). The hydrostatic stress

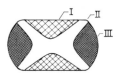

Fig. 3 Schematic illustration of the characteristic modes of deformation as seen on a longitudinal section through a cylinder upset between flat dies with no lubrication. Region I is under near-hydrostatid pressure; region II undergoes high shear; region III undergoes small axial compression and circumferential tension.

Fig. 4 Micrographs showing the characteristics of densification in 601 AB sintered aluminum alloy powder cylinders upset at 700°F (370°C): (a) no lubrication; (b) graphite in grease lubrication. (Original density 80% of theoretical.)

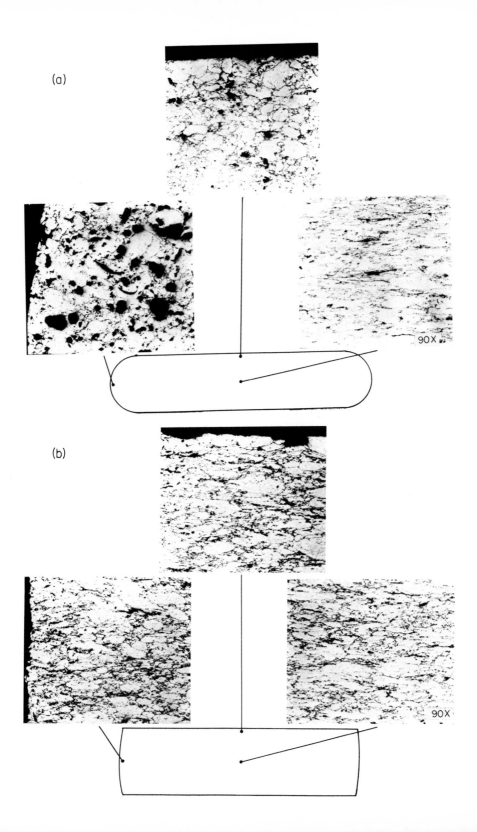

(a)

(b)

90 X

90 X

state near the die contact surfaces of Fig. 4(a) provides little densification.

The degree of nonuniformity of densification during upsetting has important practical implications. As shown in Fig. 2(c), typical powder forging processes involve upsetting deformation followed by repressing after the workpiece comes into contact with the die walls. If the upsetting deformation is very nonuniform and bulging of the expanding surface occurs (Fig. 4(a), this bulge surface will cool rapidly when it contacts the die walls. Since the bulge surface also contains large pores, they will be difficult to eliminate by the pressure of forging because cooling of this region will increase the local flow stress. For this reason, it is common to find some residual porosity at the surfaces of hot-forged powder preforms. It is clear, then, that as much densification as possible should be achieved during the upsetting mode of deformation prior to the repressing mode, preferably with well-lubricated dies (Fig. 4(b)).

B. Plastic Flow

Some details of the plastic flow of porous metals were obtained in the previous section. Determination of the fundamental aspects of plastic flow, not complicated by friction, requires a refined experimental technique. Specifically, the compression test must be performed under frictionless conditions by using a teflon sheet between the dies and specimens (Fig. 5). This provides uniform deformation, with no barreling of the outer surface, and can be used to determine the densification and plastic flow characteristics of the porous material under a pure uniaxial compressive stress. Compression of cylinders with and without friction is contrasted in Fig. 6. Frictionless die surfaces result in straight cylindrical sides, while friction leads to barreling of the free surface and consequent circumferential tensile stress. The use of this variation on the upset test for fracture will be described in Section II.C.

1. *Stress–Strain Behavior*

In compression under frictionless conditions, the applied stress is the flow stress of the material, given by

$$\sigma_z = F/A \tag{2}$$

where F is the compressive load, $A = \pi D^2/4$ the cross-sectional area, and D the cylinder diameter. The resulting compressive true strain is

$$-\epsilon_z = \ln(h_0/h) \tag{3}$$

where h_0 and h are the initial and compressed heights, respectively. Measure-

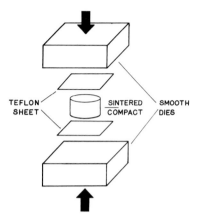

Fig. 5 Schematic exploded view of tooling for frictionless compression tests.

Fig. 6 Comparison of cylinders compressed with friction (barreled outer surface; note diagonal crack) and without friction (straight outer surface).

ments of F, D, and h after each of several increments of compressive deformation, substituted into Eqs. (2) and (3), can be used to construct the plastic true stress–true strain curve for the material.

In cold working (and hot working for small strains), a porous material will exhibit the usual strain-hardening characteristic of increasing flow stress with increasing strain. The rate of increase of stress with strain, however, will be greater than that for the same material in the fully dense condition. The continuous reduction of porosity during compression increases the fraction of the total cross-sectional area taken up by metal. This increase in the effective metal area for carrying load is called "geometric work hardening" and adds to the usual metallurgical strain hardening due to restraint of dislocation motion. This effect is evident in the log σ_z–log ϵ_z curves given in Fig. 7, which show an increasing slope, or work hardening capacity, with decreasing preform density. Smaller preform densities provide greater capacity for geometric work hardening. Fitting straight lines to the data gives the slopes of the lines, which are the exponents n in the expression $\sigma = K\epsilon^n$; n is the workhardening exponent. An empirical relation between n and preform density has been formulated;

$$n = 0.31\,\rho_0^{-1.91} \tag{4}$$

where ρ_0 is the preform density expressed as a fraction of theoretical. The n value for pure iron is 0.31, and any excess over this value for porous iron is due to geometric workhardening.

Stress–strain curves for an aluminum alloy powder at room temperature and a hot working temperature are given in Fig. 8. The cold working curve shows a high degree of workhardening, while the hot working curve, after a strain of ~ 0.1, shows very little workhardening. The increase in stress with strain that does occur in hot working at large strains is due to geometric workhardening.

2. Poisson Ratio

During axial compression of a cylinder, a tensile transverse strain accompanies the vertical compressive strain. The ratio of the transverse strain to

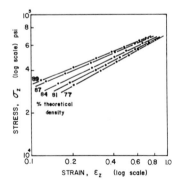

Fig. 7 Log σ_z versus log ϵ_z, for frictionless compression of sintered (2000°F) MH-100 iron powder compacts. Increasing initial compact density decreases the slope of the line (n-value). (1000 psi = 6.9 Mpa.)

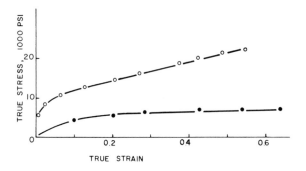

Fig. 8 Compressive true stress–true strain curves for 601 AB sintered aluminum alloy powder compacts: ○ room temperature; ● 700° F. (1000 psi = 6.9 MPa.)

axial strain is Poisson's ratio for plastic deformation, given by

$$v = -\epsilon_d/\epsilon_z = \ln(D/D_0)/\ln(h_0/h) \tag{5}$$

where $\epsilon_d = \ln(D/D_0)$ is the diametral (transverse) strain. An understanding of this effect is useful in design of dies for forging.

For a fully dense material, the Poisson ratio for plastic deformation is 0.5. This is a direct result of the fact that volume remains constant during plastic deformation. For example, equating the volume of a cylinder before and after deformation,

$$h_0(\pi D_0^2/4) = h(\pi D^2/4) \tag{6}$$

Dividing by hD_0^2 gives

$$h_0/h = (D/D_0)^2 \tag{7}$$

and taking logarithms

$$\ln(h_0/h) = \ln(D/D_0)^2 = 2\ln(D/D_0) \tag{8}$$

or

$$-\epsilon_z = 2\epsilon_d \tag{9}$$

and it is clear, from the definition of Poisson ratio (Eq. (4)), that

$$v = -\epsilon_d/\epsilon_z = 1/2 \tag{10}$$

During compressive deformation of a sintered powder metal, some material flows into the pores and there is a volume decrease. For a given reduction in height, the diameter of a powder metal cylinder will expand a lesser amount than a fully dense material. Therefore, the Poisson ratio for plastic deformation of a sintered powder material will be less than one-half and will be a function of the pore volume fraction.

Frictionless compression tests, as used for determination of the flow stress curves previously, also permit accurate measurement of the Poisson ratio. Typical results of increasing density and diametral strain with increasing height strain are given in Fig. 9. Densification and lateral flow occur simultaneously rather than in sequence.

Since the density increases during compression, the Poisson ratio also increases and is determined from the slope of the curves ϵ_d versus $-\epsilon_z$,

$$v = -d\epsilon_d/d\epsilon_z \tag{11}$$

The dashed line in Fig. 9 represents the relationship between diameter strain and height strain for a fully dense material (slope of one-half, Eq. (10)). Note that the curves for porous material are below the dashed line but gradually become parallel to it as full density is approached.

From data such as Fig. 9 and the use of Eq. (11), the variation of the Poisson ratio with density can be plotted. Figure 10 gives the results for room temperature deformation of sintered iron powder. As full density is reached, the Poisson ratio extrapolates to one-half. Note that the data follow the same curve regardless of the initial density. Similar results have been determined for room temperature deformation of copper and aluminum powders (Kuhn, 1972) and for hot deformation of aluminum alloy powder (Fig. 11). The relationship between Poisson ratio and density is

$$v = 0.5 \, \rho^a \tag{12}$$

where ρ is the fractional theoretical density. The best fit to experimental data is obtained with $a = 1.92$ for room temperature deformation and $a = 2.0$ for hot deformation. The slight difference in the exponent may be due to workhardening. It is clear that the level of porosity provides the overwhelm-

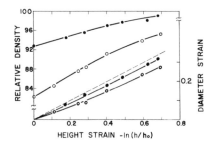

Fig. 9 Increasing density and diameter strain with increasing compressive height strain for frictionless compression of 601 AB sintered aluminum alloy powder compacts.

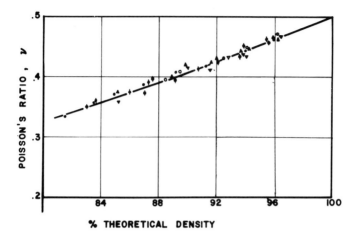

Fig. 10 Variation of plastic Poisson ratio with increasing density in cold deformation of sintered MH-100 iron powder compacts: —— $v = 0.5\rho^{1.92}$; ϕ 80%, ▼84%, ϕ 87%, ▽ 89% (1800°F); ● 80%, ▲84%, ○87%, △89% (2000°F).

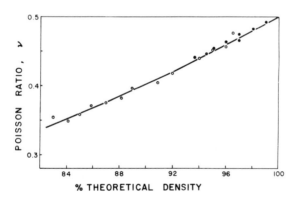

Fig. 11 Variation of plastic Poisson ratio with increasing density in hot deformation of sintered (700°F) 601 AB aluminum alloy powder compacts. Initial density: ○ 81%, ● 93%. —— $v = 0.5\rho^2$.

ing influence on Poisson ratio, with initial density and temperature having a negligible effect.

3. Plasticity Theory

In deformation processing of materials, plasticity theory is useful for calculation of forming loads and stress distributions. It is well known that plastic deformation of a fully dense material does not involve a change in volume of the material. As shown in Section II.B.2, this implies the plastic

Poisson ratio is one-half. Furthermore, the volume constancy condition is inseparable from the fact that hydrostatic stress has no influence on yield behavior of a fully dense material. (Hydrostatic stress is the condition when stress is constant in all directions, $\sigma_1 = \sigma_2 = \sigma_3$.) This is expressed in the yield criterion for a fully dense material, which is a function only of the second invariant of the stress deviator

$$f = (3J_2')^{1/2} \tag{13}$$

where f is the yield surface, J_2' the second invariant of stress deviator given by

$$[(\sigma_1 - \sigma_2)^2 + (\sigma_2 - \sigma_3)^2 + (\sigma_3 - \sigma_1)^2]/6$$

and σ_1, σ_2, σ_3 the principal stresses. It is clear that yielding depends only on stress differences and that a pure hydrostatic stress, $\sigma_h = \sigma_1 = \sigma_2 = \sigma_3$, will not influence yielding.

Sintered powder materials change volume during plastic deformation, and the Poisson ratio is, consequently, less than one-half (Figs. 10 and 11). As a further consequence, the hydrostatic stress will affect yielding. The yield behavior of other materials sensitive to hydrostatic pressure has been studied previously, such as the yielding of soils by Drucker and Prager (1952). A general yield criterion has been developed for these materials which includes the hydrostatic stress effect through the first invariant of stress

$$f = J_2'^{1/2} + \alpha J_1 \tag{14}$$

where $J_1 = \sigma_1 + \sigma_2 + \sigma_3 = 3\sigma_h$ and α is a material dependent constant. The yield criterion given in Eq. (14) includes J_1 as a linear term, which is consistent with the behavior of granular materials having limited cohesive strength. It would be expected, however, that yielding of a sintered powder material would be similar under hydrostatic tension and hydrostatic compression. The yield criterion for these materials should, then, be an even function of J_1.

A criterion proposed by Green (1972) satisfies this condition and is based on elastic–plastic analysis of deformation of a volume element containing a spherical pore. The criterion is given by

$$\delta Y^2 = J_2' + \alpha J_1^2 \tag{15}$$

where α and δ are functions of porosity and Y is the yield strength of the solid material. Deformation equations resulting from this criterion give the Poisson ratio as

$$v = (1 - 2\alpha)/(2 + 2\alpha)^2 \tag{16}$$

where $\alpha = 0.75 \, (1 - v^{1/3})/(3 - 2v^{1/4}) \ln v$ and $v = 1 - \rho$. The relationship

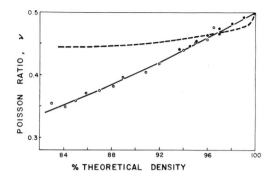

Fig. 12 Comparison of predicted v versus ρ relationship due to Green (1972) with experimental relationship of Fig. 11. – – – Green, Eq. (16).

given by Eq. (16) is represented in Fig. 12 by the dashed line. It is clear that this relationship overestimates the experimental measurements of the Poisson ratio at low densities and predicts a much lower sensitivity to increasing density.

Another criterion of yielding for porous material proposed by Kuhn and Downey (1971) is based on the observed relationship between the Poisson ratio and density (Eq. (12)) and a modification of the yield criterion (Eq. (13)). The criterion is given by

$$f = \left[3J_2' - (1 - 2v) J_2 \right]^{1/2} \tag{17}$$

where J_2 is the second invariant of stress, given by $J_2' - J_1^2/3$. Hydrostatic stress is included as a quadratic term through J_2 and thus satisfies the requirement that the yield criterion be an even function of hydrostatic stress. Since the second term in Eq. (17) includes the factor $(1 - 2v)$, the effect of J_2 (hence, J_1) will vanish as $v \to 0.5$ or as full density is reached (see Eq. (12)). Thus, the yield criterion of Eq. (17) reduces to that for fully dense material (Eq. (13)) as $\rho \to 1.0$. Yield surfaces defined by Eq. (17) are illustrated in Fig. 13.

If the normality flow rule is assumed to hold for plastic flow of a porous material, the plastic strain increments can be derived from Eq. (17) to give

$$d\epsilon_1 = d\lambda \left[\sigma_1 - v(\sigma_2 + \sigma_3) \right]$$
$$d\epsilon_2 = d\lambda \left[\sigma_2 - v(\sigma_3 + \sigma_1) \right] \tag{18}$$
$$d\epsilon_3 = d\lambda \left[\sigma_3 - v(\sigma_1 + \sigma_2) \right]$$

where $d\lambda = d\bar{\epsilon}/\bar{\sigma}$. The effective strain increment and effective stress are

$$d\bar{\epsilon} = \{E_2' + (1 - 2v)[v(2 - v)E_2' - E_2]\}^{1/2} \tag{19}$$

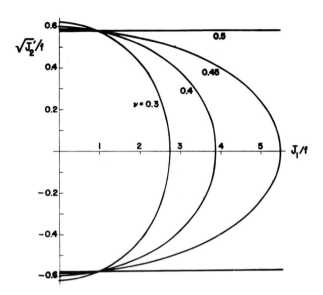

Fig. 13 Yield surfaces for the yield criterion of Eq. (17) (Kuhn and Downey, 1971). As density approaches 1.0 (Poisson ratio approaches 0.5), the yield surface approaches the von Mises cylindrical yield surface.

where

$$E_2' = \left[(d\epsilon_1 - d\epsilon_2)^2 + (d\epsilon_2 - d\epsilon_3)^2 + (d\epsilon_3 - d\epsilon_1)^2\right]/2(1 + v)^2$$
$$E_2 = -(d\epsilon_1 d\epsilon_2 + d\epsilon_2 d\epsilon_3 + d\epsilon_3 d\epsilon_1)$$

and

$$\bar{\sigma} = \left[3J_2' - (1 - 2v)J_2\right]^{1/2} \tag{20}$$

For frictionless uniaxial compression, the applied strain increment is $d\epsilon_1$, and the transverse strains are $d\epsilon_2 = d\epsilon_3 = -vd\epsilon_1$. In this case, $d\bar{\epsilon} = d\epsilon_1$ from Eq. (19). Likewise, the applied stress is $\sigma_1 = Y$, the flow stress of the material, with $\sigma_2 = \sigma_3 = 0$. From Eq. (20), $\sigma_1 = Y = \bar{\sigma}(\bar{\epsilon})$, where the effective stress is a function of total effective strain $\bar{\epsilon}$, given by $\bar{\epsilon} = \int d\bar{\epsilon}$. The stress–strain curve for the material can thus be determined from a frictionless uniaxial compression test (e.g., Figs. 7 and 8).

Densification during plastic deformation is best determined from

$$d\rho/\rho = -(d\epsilon_1 + d\epsilon_2 + d\epsilon_3) \tag{21}$$

where $d\rho$ is the incremental change in density due to incremental strains $d\epsilon_1$, $d\epsilon_2$, and $d\epsilon_3$. Caution should be exercised in using the popular expression for volume (or density) change

$$\Delta V/V = -\Delta\rho/\rho = \epsilon_1 + \epsilon_2 + \epsilon_3 \qquad (22)$$

where ϵ_1, ϵ_2, and ϵ_3 are total true strains. This relationship is valid for small volume (or density) changes, as in plastic deformation of a fully dense material, in which $\Delta V/V = 0$. In plastic deformation of sintered powder metal, however, the density change may be as large as 30%. In this case, the correct expression for volume change is

$$\ln(1 + \Delta V/V) = \ln(1 - \Delta\rho/\rho) = \epsilon_1 + \epsilon_2 + \epsilon_3 \qquad (23)$$

and the error in using Eq. (22) is substantial. This expression is inconvenient for analysis of plastic deformation, and Eq. (21) is used in the following analyses. Equations (18)–(21), along with Eq. (12), constitute a complete set of relationships for the calculation of forming stresses and densification in deformation processing of sintered powder materials.

C. Fracture

Fracture during plastic deformation of fully dense materials initiates by void formation at inclusions or other structural inhomogeneities. These voids coalesce along planes of maximum shear to form the complete fracture. The important role of inhomogeneities is clearly illustrated in Fig. 14, showing the reduced tensile ductility with increasing volume fraction of second-phase particles (Edelson and Baldwin, 1962).

Preexistence of voids in material eliminates the need for the first step in the ductile fracture process and only coalescence of voids is necessary. The detrimental effect of voids on ductility in sintered powder materials is shown clearly by Kaufman and Mocarski (1971) (see Chapter 5). Thus, material having a given composition and inclusion content will experience a further decrease in ductility if another phase consisting of voids is introduced. For this reason, workability of sintered powder materials is severely limited.

In deformation processing of materials, tensile ductility is not always a true indicator of the ability of a material to be deformed. During processing the material is generally subjected to combined stresses and understanding of fracture under such stress states is required. Although it is well known that deformation to fracture is enhanced by hydrostatic pressure (Bridgman, 1944; Coffin and Rogers, 1967), the quantitative representation of this effect is necessary for use in deformation processing.

One useful technique that has been developed for evaluation of deformation to fracture (workability) under conditions of deformation processing is the upset test. During axial compression of a cylinder, friction at the die

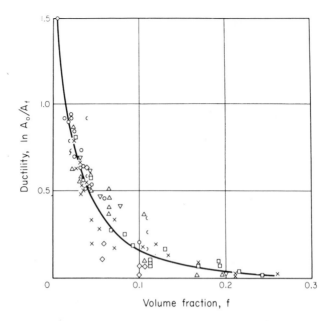

Fig. 14 Decrease in tensile ductility with increasing fraction of second phase particles or holes. (Edelson and Baldwin, 1962.) Present work: ◇ copper–iron–molybdenum, ○ copper holes, X copper chromium, □ copper alumina, △ copper iron, ▽ copper molybdenum. Zwilsky and Grant: ⊂ copper alumina, ⊃ copper silica.

contact surfaces retards radial outward flow of material at these surfaces and leads to barreling of the cylindrical free surface, as shown in Figs. 3, 4, and 6. Curvature of the bulge surface increases with increasing friction and decreasing aspect ratio (H/D, height-to-diameter ratio). Under frictionless compression, the cylindrical free surface remains straight and the deformation is homogeneous with no tensile stress in the circumferential direction. The nonuniform deformation under conditions with friction, however, results in a secondary tensile stress in the circumferential direction accompanying the axial compressive stress (Fig. 15). Increasing barreling curvature due to increasing friction or decreasing H/D increases the circumferential tensile stress. Since fractures occur as a result of this tensile stress, the upset test may be utilized to study fracture during deformation processing of sintered powder materials.

As an illustration, Fig. 16 gives the overall deformation at fracture in compression of cylinders of 601 AB aluminum alloy powder. It is clear that increasing the aspect ratio (H_0/D_0) increases the deformation to fracture, as does the use of a lubricant to reduce die contact friction. A surprising result is that deformation to fracture is essentially independent of original

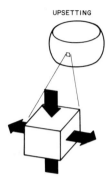

Fig. 15 Circumferential tensile stress and axial compressive stress at the bulge surface of cylinders compressed axially with high contact friction.

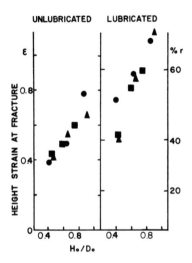

Fig. 16 Overall height strain at fraction for hot upsetting of sintered (700°F) 601 AB aluminum alloy powder compacts: ■ 76%, ● 88%, △93%.

porosity in the material. This apparent anomaly can be explained if it is first recalled that fracture is the result of interaction between local stress state and microstructure of the material. As initial pore volume increases, the ability of the material to withstand tensile stresses without fracture decreases. Concurrently, the increase in initial pore volume results in a smaller Poisson ratio and decrease in degree of lateral spread. Thus, the tendency for bulging and attendant circumferential tensile stress will decrease. The net result of decreased local ductility and decreased tensile stress, both due to increased porosity, is no change in overall compressive deformation to fracture.

Further measurements on the same upset tests provide results for establishment of a fracture criterion for porous metals. The technique was developed originally for fully dense material by Kuhn *et al.* (1973) and Lee and Kuhn (1973). Measurements are made of the axial and circumferential strains on small grid marks at the equator of the bulge surface of the upset cylinders, depicted in Fig. 16. The ratio of these strains during compression is altered by modifying the friction condition and aspect ratio of the cylinder. At fracture, the surface strains fit a straight line having a slope of one-half on tensile strain–compressive strain axes. Figure 17 gives the results of such tests for 601 AB aluminum alloy powder and Fig. 18 for 4620 low alloy steel powder.

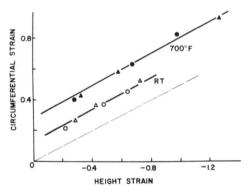

Fig. 17 Locus of surface strains at fracture during upsetting of sintered aluminum allow powder cylinders: ○, ● 201 AB; △, ▲ 601 AB (RT = room temperature).

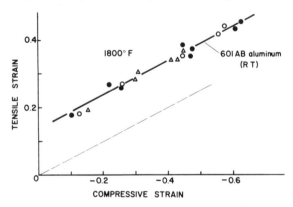

Fig. 18 Locus of surface strains at fracture during upsetting of sintered 4620 low alloy steel powder cylinders. Induction sinter: ● 2350°F (3 min), ● 2050°F (3 min). Conventional sinter: △ 2050°F (0.5 hr).

Fracture strain loci, such as Figs. 17 and 18, can be considered as fracture criteria for evaluation of the deformations to fracture in more complex deformation processes. The progressing deformation strain paths in potential fracture regions of the process under consideration are first determined through plasticity analysis or measurements on a model material. These strain paths are then compared with the fracture locus of the material. Any strain path crossing the fracture locus before the deformation process is complete indicates that fracture is likely and alterations of the process may be utilized to change the strain paths so they do not cross the fracture locus.

The fracture lines in Figs. 17 and 18 give a locus of fracture strains for free surface fracture. Other types of fracture in metalworking processes are die–contact–surface cracks and internal defects. These are discussed in the particular case of extrusion forging in Section III.B.

III. APPLICATIONS

The fundamental relationships of porous metal densification, plasticity, and fracture may be used for design of deformation processes. Application of the results first requires examination of several basic modes of metal flow that occur during forging of complex parts. In this section, the characteristics of compressive deformation and extrusion flow are presented, followed by application of the results to some complex shapes.

A. Compression

In forging of sintered powder preforms, several types of compressive deformation may occur, including simple upsetting (Fig. 2(c)), plane strain compression, and repressing (Fig. 2(b)). Densification, pressure, and metal flow during each of these types of deformation are illustrated.

1. Upsetting

In frictionless compression tests, which were used to determine the fundamental plastic flow behavior of porous metals in Section II.B, the deformation is uniform with no barreling of the cylindrical surface. Taking the applied axial compressive strain increment as $d\epsilon_z$, the transverse strains, making use of Eq. (12), are

$$d\epsilon_\theta = d\epsilon_r = -\nu d\epsilon_z = -0.5\rho^2 d\epsilon_z \tag{24}$$

Substituting these strains into Eq. (21), the density change is given by

$$d\rho/\rho = (1 - \rho^2)d\epsilon_z \tag{25}$$

Integration gives the relation between axial compressive strain and density

$$-\epsilon_z = \ln\left[(\rho/\rho_0)^2(1 - \rho_0^2)/(1 - \rho^2)\right]^{1/2} \tag{26}$$

where ρ is the relative density of the deformed cylinder and ρ_0 is the initial relative density (preform density).

Calculations from Eq. (26) are compared in Fig. 19 with experimental results obtained by Antes (1971) for compression of cylindrical compacts of atomized iron and 1020 and 4620 steel powders. The correlation shown in Fig. 19 indicates the universal applicability of the Poisson ratio/density relationship (Eq.(12)).

2. Plane Strain Compression

In plane strain deformation, depicted in Fig. 20, compression in the z direction results in transverse strain in the y direction, but strain in the x direction is zero. Constraint of flow in the x direction could be the result of contact with die sidewalls, or due to adjoining undeformed material, as

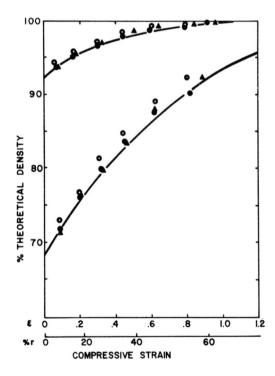

Fig. 19 Comparison of calculated density, Eq. (26), with experimental data from Antes (1971) for uniaxial compression: ▲ atomized iron, ● 1020, ○ 4620, —— calculated.

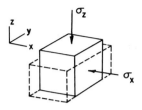

Fig. 20   Schematic illustration of plane strain deformation.

in the plane strain compression test. In any case, a stress σ_x is developed in the x direction. Stress in the y direction is zero. Taking x, y, z equivalent to 1, 2, 3 in Eqs. (18), the condition $d\epsilon_x = 0$ gives

$$\sigma_x = v\sigma_z \tag{27}$$

and, in turn,

$$d\epsilon_y = -v/(1 - v)d\epsilon_z \tag{28}$$

The density change, from Eq. (21) and the Poisson ratio/density relationship (Eq. (12)), is

$$-d\rho/\rho = d\epsilon_z(1 - \rho^2)/(1 - 0.5\rho^2) \tag{29}$$

Numerical integration of Eq. (29) gives the relation between density and height strain in plane strain compression. A comparison of this calculation with experimental results from Antes (1971) is shown in Fig. 21. The correlation again indicates the validity of the Poisson ratio/density relationship and the proposed plasticity theory (Eqs. (18)–(21)).

The applied stress in plane strain compression is determined by substitution of $\sigma_y = 0$ and $\sigma_x = v\sigma_z$ into the yield criterion, or effective stress, Eq. (20), giving

$$\sigma_z = \sigma(\bar{\epsilon})/(1 - 0.25\,\rho^4)^{1/2} \tag{30}$$

Substituting $d\epsilon_x = 0$ and $d\epsilon_y = -v/(1 - v)d\epsilon_z$ into Eq. (19) gives the equivalent strain increment; utilizing Eq. (29), the result is

$$d\bar{\epsilon} = \left[(1 - 0.5\rho^2)/(1 - \rho^2)(1 - 0.25\rho^4)^{1/2}\right]d\rho \tag{31}$$

Then $\bar{\epsilon}$ is found from $\bar{\epsilon} = \int d\bar{\epsilon}$, and $\bar{\sigma}(\bar{\epsilon})$ is found from the stress–strain curve for the material. Using this method, the forming stress σ_z in plane strain compression was calculated from Eq. (30). Comparison of theoretical results with experimental data from Antes (1971) shows good agreement (Kuhn, 1972).

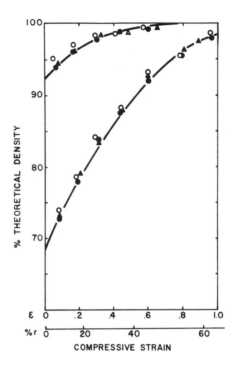

Fig. 21 Comparison of calculated density, Eq. (29), with experimental data from Antes (1971) for plane strain compression: ▲ atomized iron, ● 1020, ○ 4620, ——— calculated.

3. Repressing

In compression of a preform within a completely closed die (repressing), the material is in contact with the die sidewalls throughout the deformation process. Repressing is unique to powder metallurgy and is used to increase the density of conventional sintered powdered metal parts. Constraint by the die sidewalls produces a transverse stress, as depicted in Fig. 22. The applied vertical strain increment is $d\epsilon_z$ and the strain components $d\epsilon_r$ and

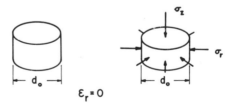

Fig. 22 Schematic illustration of repressing deformation.

$d\epsilon_\theta$ are zero. Then the effective strain increment, from Eq. (19), is

$$d\bar{\epsilon} = d\epsilon_z[1 + \rho^2(1 - 0.25\rho^2)(1 - \rho)]^{1/2}/(1 + 0.5\rho^2) \tag{32}$$

and the density increment is simply $d\rho/\rho = -d\epsilon_z$. For this case, Eqs. (18) give

$$\sigma_r = \sigma_\theta = \sigma_z v/(1 - v) \tag{33}$$

and the forming stress is found from Eq. (20),

$$\sigma_z = \bar{\sigma}(\bar{\epsilon})[(1 - 0.5\rho^2)/(1 - \rho^2)(1 + 0.5\rho^2)] \tag{34}$$

The total effective strain is determined from $\bar{\epsilon} = \int d\bar{\epsilon}$, and $\bar{\sigma}(\bar{\epsilon})$ is found from the stress–strain curve for the material.

Calculations using Eq. (34) are compared in Fig. 23 with experimental data on repressing of 1020 steel powder from Antes (1971). The good correlation again indicates the validity of the proposed plasticity theory for porous metals. Similar correlations are obtained in comparison of calculated results with data on hot repressing of steel powder by Bockstiegel and Olsen (1971).

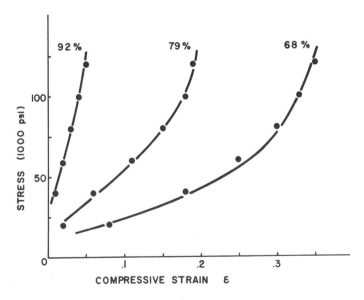

Fig. 23 Comparison of calculated pressure with experimental data from Antes (1971) for repressing: ● 1020 steel, —— calculated.

4. Compression with Friction

The previous analyses of compressive deformation neglected the effects of friction at the die contact surfaces. An estimate of the influence of friction on densification and forming stress was determined for axisymmetric compression through use of the slab method of analysis (Kuhn and Downey, 1973).

Die contact surface friction is assumed to follow the law

$$\tau = mk \tag{35}$$

where k is the shear yield strength of the material and m is the friction factor $(0 \leq m \leq 1)$. Referring to Fig. 24, equilibrium gives the expression for transverse pressure

$$q = 2mk(R - r)H \tag{36}$$

where R is the outside radius, H the height, and q the transverse pressure at point r.

Assuming the circumferential stress is equal to the radial stress, the effective stress, Eq. (20), becomes

$$\bar{\sigma}(\bar{\epsilon}) = \left[(p - q)^2 + (1 + 2v)(q^2 + 2pq)\right]^{1/2} \tag{37}$$

where p is the vertical pressure. Solving for p and normalizing with respect to $\bar{\sigma}$,

$$p/\bar{\sigma} = (2vq/\bar{\sigma}) + \left[1 - 2(q/\bar{\sigma})^2(1 - v - 2v^2)\right]^{1/2} \tag{38}$$

From Eqs. (18), the ratio of radial strain increment to axial strain incre-

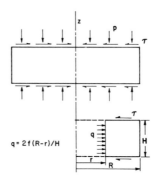

$$q = 2f(R-r)/H$$

Fig. 24 Internal pressure q due to compression of a cylinder with friction at the die surface.

ment is

$$-d\epsilon_r/d\epsilon_z = [q(1 - v) - vp]/(2vq - p) \tag{39}$$

Substituting Eqs. (36) and (38) gives

$$-d\epsilon_r/d\epsilon_z = \rho^2/2 - MB(1 - r/R)/[1 - MB^2(1 - r/R)^2B]^{1/2} \tag{40}$$

where $M = mR/H$ and $B = 1 - \rho^2/2 - \rho^4/2$. Finally, the fractional change in density is given by

$$-d\rho/\rho = d\epsilon_r + d\epsilon_\theta + d\epsilon_z = d\epsilon_z(1 + 2d\epsilon_r/d\epsilon_z) \tag{41}$$

Substituting Eq. (40), the increment in density with increment in axial strain $d\rho/d\epsilon_z$ can be calculated.

Figure 25 gives the calculated results of $d\rho/d\epsilon_z$ versus r for three values of m and preform density of 80%. Note that, for frictionless conditions, the density increment is uniform across the sample. For increasing values of friction, however, the density increment increases toward the center of the cylinder. This is the result of increased transverse pressure at the interior, as indicated in Fig. 24. Further calculations of average density increase

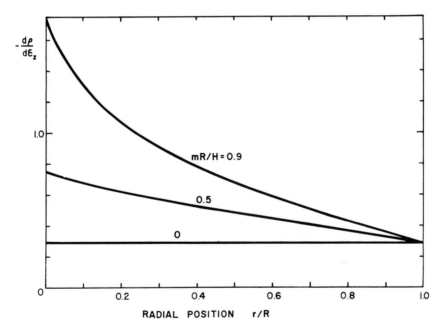

Fig. 25 Calculated change in density with axial compression, Eqs. (40) and (41), showing the effect of die contact friction. Preform density = 80%.

from Eq. (41) showed good correlation with experimental measurements (Kuhn and Downey, 1971).

5. Design of Disk Forging

In Section II.A.2, it was inferred, and in the next chapter it will be shown, that at least 50% height reduction is required to achieve maximum properties in forged steel powder. It was also shown in Section II.C that fracture limits the amount of compression that can be imposed on a material before fracture occurs (Fig. 16). Design of preforms for powder forging must be such that sufficient metal flow occurs to achieve maximum properties, but less than that at which fracture occurs. A design technique for achieving these objectives is demonstrated for the process of forging cylindrical preforms into disks.

For a given disk diameter and height, it is required to specify the cylindrical preform geometry (height and diameter) and forging process conditions that lead to sufficient deformation for maximum (or required) properties without the occurrence of fracture during forging. Fracture in upsetting (e.g., in Fig. 6) Can be avoided if the expanding free surface of the cylinder reaches the die sidewalls before the reduction is reached at which fracture would otherwise occur (see Fig. 2(c)). This prevents further development of circumferential tensile stresses, depicted in Fig. 15.

As an example, consider a material requiring 50% reduction in height to reach maximum properties (see Chapter 5). Equating the mass of the preform to the mass of the forged disk, the required disk aspect ratio (height-to-diameter ratio) is

$$H_f/D_f = (H_0/D_0)\rho_0^{-1/2}(H_f/H_0)^{3/2} \tag{42}$$

where H_0, D_0, and ρ_0 are the height, diameter, and relative density, respectively, of the preform.

Similarly, if the material exhibits a relation between height strain at fracture and preform aspect ratio H_0/D_0 as in Fig. 16, then the forged disk aspect ratio at fracture is

$$H_f/D_f = (H_0/D_0)\rho_0^{-1/2}\exp[-3/2\epsilon_H] \tag{43}$$

where ϵ_H is determined, for a particular preform aspect ratio H_0/D_0, from Fig. 16.

Equations (42) and (43) are plotted in Fig. 26 on axes of preform aspect ratio versus forged disk aspect ratio. For a given preform aspect ratio, the forged disk aspect ratio must be to the left of the solid line to achieve the 50% reduction required for maximum properties (i.e., $H_f/H_0 = 0.5$ in Eq. (42)). In addition, the forged disk aspect ratio must be to the right

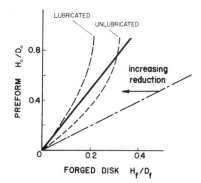

Fig. 26 Design curves for disk forging to achieve maximum mechanical properties (left of solid line) without causing fracture during forming (right of dashed line).

of the dashed line in order to avoid fracture. In other words, starting at the reference line for a given H_0/D_0, move horizontally to the left as deformation proceeds. For maximum properties, we must cross the solid line but avoid crossing the dashed line to prevent fracture. It is clear that, for the unlubricated case, forged disk H_f/D_f must be greater than 0.3, and the corresponding preform H_0/D_0 must be greater than 0.78. For the lubricated case, any forged disk aspect ratio can be formed, but there is a narrow corridor of allowable preform aspect ratios for successful forging. These limits will be different, of course, for materials having characteristics other than those in Fig. 16.

B. Extrusion Forging

Many forged parts contain hubs and rims, which are formed by extrusion. The deformation in this case is more complex than in the simple compression modes of deformation described in Section III.A, and do not lend themselves to simple plasticity analysis. The major limitation in the extrusion models of deformation is fracture. These limits, and the effects of preform and process parameters have been determined experimentally.

Extrusion of a hub is illustrated in Fig. 27. During deformation, fractures may occur on the top surface due to tensile strains arising from friction along the die surfaces. The top is a free surface until it contacts the flat section of the die. Cracking may also occur along the sides of the hub at points in contact with the die. Finally, if two opposed hubs are formed, an internal fracture may occur at the center of the part.

1. Free Surface Fracture

As illustrated in Fig. 27, extrusion forging of a hub involves compression of the flange section with radial flow inward and up into the hub sections. The top surface is a free surface that undergoes bulging and tensile strains due to friction along the sides of the die. These strains are nominally equal and the strain path extends into the first quadrant of strain space (Fig. 27). Fracture occurs on the top surface when the strains reach the fracture line.

Typically, for small draft angles, the hubs reach a height roughly equal to the hub base diameter before fracture occurs. Use of large draft angles (>40°) drastically reduces the hub height at fracture. Increasing friction or decreasing the die corner radius increases the tensile strains at the top surface and, thus, decreases the hub height at fracture.

An alternative approach involves use of a preform that partially fills the hub section of the die. Then metal flow occurs simultaneously upward by extrusion and radially outward by upsetting (lateral flow). The nature of the strains on the top surface, however, depends on the die angle and friction, as illustrated in Fig. 28. For a friction coefficient of 0.1, draft angles of 40° and 50° lead to tensile strains at the hub top surface, and fracture occurs as the strain paths cross the fracture line, as in Fig. 27. Hub heights at fracture are very small in this case. For die angles 10°, 20°, and 30°, the strains on the top free surface are compressive initially, then reverse and move toward the origin. Decreasing die angle increases the compressive strain to reversal. The reversal occurs because friction along the conical die surfaces overcomes the compressive radial strains resulting from the decrease in diameter of the top surface. Fracture then occurs when the tensile strain after

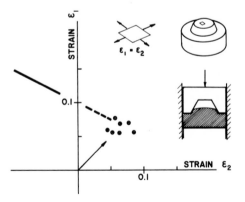

Fig. 27 Strains at fracture at the top free surface in hub extrusion of sintered 601 AB aluminum alloy powder.

the strain path reversal is approximately equal to the tensile strain to fracture in the first quadrant. Fracture did not occur in the 10° die since the hub section reached the top of the die before the strains became large enough for fracture. For the small die angles, hub heights reach approximately twice the hub base diameter before fracture occurs.

For any given friction coefficient, there is a specific draft angle above which the hub top surface strains are tensile and below which the strains are initially compressive. These angles have been predicted through slip-line field analysis and confirmed experimentally by Suh *et al.* (1976).

2. Die Contact Surface Fracture

Frequently, cracks occur during forging on workpiece surfaces that are in contact with the die. One common location of such defects is the vicinity of a die or punch corner. Figure 29 illustrates the initiation of such a crack at the die corner and its migration up the conical surface of the die during hub extrusion. Fundamental studies by Suh and Kuhn (1977) show that a

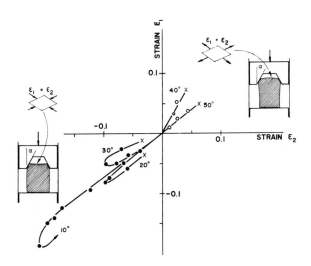

Fig. 28 Strain paths and strains at fracture for combined hub extrusion and radial flow: ○ upsetting mode, ● extrusion mode.

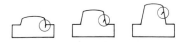

Fig. 29 Schematic illustration of initiation and migration of a die–contact–surface crack during hub extrusion.

concentration of shear deformation combined with a low level of pressure perpendicular to the shear line leads to the initiation of this type of crack.

In the present case, use of a larger die corner radius (>2 mm) eliminated the source of the problem. For small die corner radii, decreasing die angle or decreasing friction on the hub surfaces increases the likelihood of fracture at the die corner. Increasing friction or the hub draft angle increases the back pressure in the region of the die corner and helps prevent crack formation. Unlike most other types of fracture, then, increasing friction on the hub surface tends to reduce the occurrence of die contact surface cracks. Fundamental studies by Suh and Kuhn (1977) have also shown that die contact surface fractures result from nonuniform lubrication; in fact, high friction due to poor lubrication, as long as it is uniform over the contact surface, prevents contact surface fractures.

3. Internal Fracture

Internal defects, often called central burst, are a third form of fracture occurring in metalworking processes. Unlike the other types of fracture, internal fractures can remain undetected because they rarely penetrate to the surface. Although internal fractures may be closed up in subsequent deformation of the part, there is no certainty that the bond will be sound. Service failures due to internal defects are particularly catastrophic because they cannot be detected in their early stages of growth by surface inspection. These aspects emphasize the need for investigation of internal fracture, even though its occurrence is much less frequent than other types of fracture.

Internal fracture takes the form of a large crack at the center of a forged part, as shown in Fig. 30. Axial tension at the central point, caused by

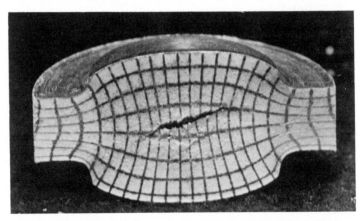

Fig. 30 Central burst in a forged part consisting of two opposed hubs. Grid lines placed on midplane prior to deformation to permit measurement of internal strains.

opposed axial flow from the center toward each hub, leads to such defects. The magnitude of tension at the center depends on the ratio of preform aspect ratio H_0/D_0. Friction has very little effect on this type of fracture. Experimental study by Suh and Kuhn (1977) on 601 AB sintered aluminum powder indicates the amount of deformation possible before internal crack formation (Fig. 31). A minimum amount of deformation to fracture occurs for $H/D = 0.35$, below which a double bulge forms on the top hub surface and above which a single bulge forms. No theoretical explanation of this result has yet been developed. Also shown on Fig. 31 are the forming limits due to surface cracks at the hub top surface (Section III.B.1) and due to buckling which occurs in thin preforms.

C. Prototype Parts

Actual forged parts have more complex configurations than the compression or extrusion modes of flow depicted in Sections III.A and III.B. Nevertheless, the complex parts can be subdivided into regions of simple deformation. Two parts are examined to illustrate this technique and demonstrate application of elementary deformation behavior to design of the proper preforms.

The first part is a flange–hub combination, shown in Fig. 32, examined by Ferguson *et al.* (1977). One approach to formation of this part is repressing, i.e., use of a preform having the same shape as the final part which is then

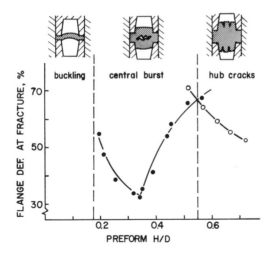

Fig. 31 Limiting deformations for buckling, central burst, and hub cracking in double hub extrusion of sintered 601 AB aluminum alloy powder.

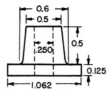

Fig. 32 Flange–hub part examined for preform design. Dimensions are in inches (1 mm = 0.04 in).

simply compressed axially for densification. Since there is no lateral flow in this approach, maximum mechanical properties are not achieved (see Chapter 5).

Alternative preforms which do involve substantial lateral flow are shown in Fig. 33. In Fig. 33(a), the preform fills the hub section, and the flange is formed by upsetting, while the preform in Fig. 33(b) fills the flange, and the hub is formed by extrusion. In the first case, the diametral strain required (refer to Fig. 32) is $\epsilon_D = \ln(1.066/0.6) = 0.575$, which is clearly in excess of the limiting strains at fracture for most powder materials (Figs. 17 and 18). In the second case, material flows upward into the hub with friction acting along the hub surfaces and the mandrel at the center. This is similar to hub extrusion, as described in Section III.B.1, where friction along the surfaces resulted in tensile stresses at the hub top surface and fracture (Fig. 27). In the present case, however, a clearance may be used between the mandrel and the preform bore diameter to reduce some of the frictional shear traction along the hub surfaces. A bore diameter

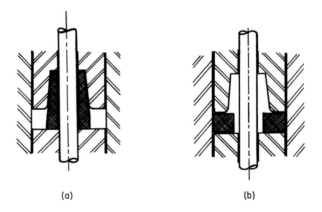

(a)   (b)

Fig. 33 Two possible preform designs for the part shown in Fig. 32.

midway between the mandrel and hub base diameters was found to provide successful deformation and densification without fracture.

A second part with added complexity, a hub–flange–rim combination (Fig. 34) was examined by Downey and Kuhn (1975). Again, the repressing alternative is rejected because of its limited development of mechanical properties. Flat ring preforms provide large amounts of metal flow in reaching the final shape.

Even though the general ring shape is specified, its dimensions must be determined. As shown in Fig. 35, the preform may have:

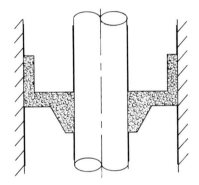

Fig. 34 Flange–hub–rim part examined for preform design.

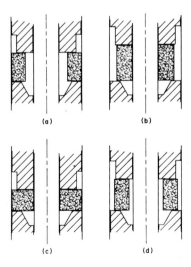

Fig. 35 Possible preform designs for the part shown in Fig. 34.

(a)   clearance at the bore diameter with no clearance at the outside diameter;

(b)   clearance at the outside but no clearance at the bore;

(c)   no clearance at the outside or the bore; and

(d)   clearance at both the outside and the bore.

Both preforms (b) and (d) can be rejected because cracks will occur at the outer rim as metal flows around the upper punch radius. This occurs because the metal is expanding in diameter as it flows around the corner, but there is axial compression to help compensate for the circumferential tension. This type of cracking can be avoided by using a preform which fills the die, i.e., has no clearance at the outside diameter, as in preforms (a) and (c).

Preform (c) can be rejected because it is similar to hub extrusion, and may lead to cracking at the hub top surface (Fig. 27). As in the preform illustrated in Fig. 33(b), use of a clearance between the bore diameter of the preform and the mandrel eliminates this type of crack. Therefore, the preform shown in Fig. 35(a) is selected for the flange–hub–rim part. Use of this preform has led to successful, defect-free parts, while the expected cracking occurred with use of the other preforms (Downey and Kuhn, 1975).

D.   Production Parts

The workability analysis concept described in Section II.C and demonstrated in Section III.C has also been used in preform design for actual parts currently formed by powder forging. Two examples are given.

One application involves cold forging ball bearing races from sintered 4620 steel powder preforms (shown in Fig. 36). Use of a beveled preform (Fig. 37) led to cracking due to high shear stress at the point of contact between the upper punch and preform. This type of defect can be avoided by using a preform which initially fills the upper rim; then during forging, material flows downward, around the punch radius, and inward toward the inner mandrel (Fig. 38(a)). In this case, the free surface on which cracking might be expected is the inner cylindrical surface. Measurement of the strains on this surface indicate that the circumferential strain (ϵ_1) is compressive since the bore diameter decreases during forging. The axial strain (ϵ_2) is also compressive initially and then becomes slightly tensile (Fig. 38(b)). Nevertheless, the strain path does not cross the fracture limit line for 4620 steel (heavy line in Fig. 38(b)) for both lubricated and unlubricated conditions. Thus, fracture does not occur in the preform shown in Fig. 38(a). Further details on production of this bearing race are given by Ferguson (1977).

A second application involves the pinion gear shown in Fig. 39. The

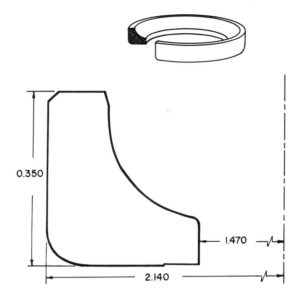

Fig. 36 Ball bearing outer race considered for powder forging. Dimensions are in inches (1 mm = 0.04 in.).

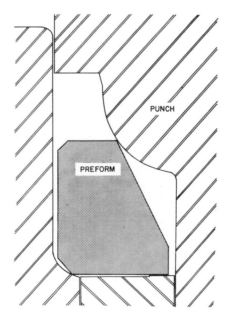

Fig. 37 Original preform concept, which led to cracking at the point of contact with the upper punch.

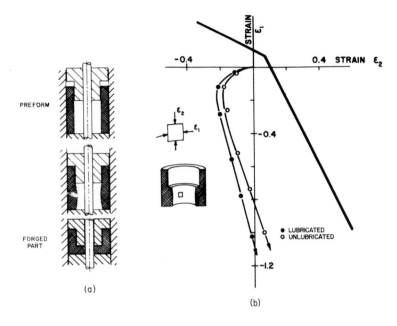

(a)

(b)

Fig. 38 (a) Modified preform leading to defect-free bearing races. (b) Strain paths on the bore surface, illustrating that they do not cross the fracture line (heavy line), and fracture does not occur.

Fig. 39 Partially formed pinion gear showing cracks forming on the teeth.

partially formed gear shows the development of cracks on the teeth as metal flows outward into the tooth sections of the die. Although the cracks close up as the material is pressed against the die surfaces during the final stages

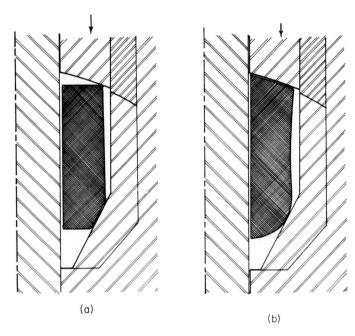

Fig. 40 Longitudinal section view of preforms and tooling. (a) Original preform, which led to crack formation at the teeth. (b) Modified preform, which led to defect-free gears.

of the forging stroke, entrapped lubricant and oxidation in the cracks retain structural weaknesses at the tooth surfaces. Original preform design involved a flat-topped preform with a beveled bottom surface (Fig. 40(a)). Modification of this design involved a slightly tapered top surface to match the contour of the upper punch (i.e., the spherical back face of the gear). Another modification involved a rounded, instead of beveled, bottom surface (Fig. 40(b)). The preform was produced by isostatic compaction. Use of this type of preform prevented cracking because contact of the preform along the entire upper punch face produces greater axial compressive strains at the free surface. These compressive strains then permit the large circumferential tensile strains at the tooth surfaces without fracture.

IV. SUMMARY AND CONCLUSIONS

The complicating effects of porosity during deformation processing of sintered powder materials are limited workability, densification, and necessity for elimination of porosity for property requirements. A workability analysis method, based on an experimental fracture strain locus for each material, has been developed to permit preform design for enhanced deformation to fracture. Use of this concept for development of preform design guidelines was demonstrated.

Densification, flow and pressures during plastic deformation of sintered powder materials have been calculated using a proposed plasticity theory for porous material. Correlation with experimental results confirms validity of the plasticity theory.

The effects of residual porosity and the characteristics of metal flow on mechanical properties of forged, sintered powder materials are treated in Chapter 5. In general, deformation involving large amounts of lateral flow and shear enhance mechanical properties of the densified powder material.

Basically, then, preform design for powder forging involves achieving a delicate balance of sufficient deformation for maximum properties, but less than the amount of deformation leading to fracture.

REFERENCES

Antes, H. W. (1971). *In* "Modern Developments in Powder Metallurgy" (H. H. Hausner ed.), Vol. 4. Plenum Press, New York.

Bockstiegel, G., and Olsen, H. (1971). *Powder Met.* **15**, 127–149.

Bridgman, P. W. (1944). *Trans. ASM* **32**, 553.

Coffin, L. F., Jr., and Rogers, H. C. (1967). *Trans ASM* **60**, 672–686.

Downey, C. L., and Kuhn, H. A. (1975). *Trans. ASME* **97H**, 121–125.

Drucker, D. C., and Prager, W. (1952). *Quart. Appl. Math.* **10**, 157.

Edelson, B. I., and Baldwin, M. M. (1962). *Trans. ASM* **55**, 230–250.

Ferguson, H. (1977). *In* "Modern Developments in Powder Metallurgy" (P. W. Taubenblat and H. H. Hausner, eds.), Vol. 9. Metal Powder Industries Federation, Princeton, New Jersey.

Ferguson, B. L., Kuhn, H. A., and Lawley, A. (1977). *In* "Modern Developments in Powder Metallurgy" (P. W. Taubenblat and H. H. Hausner, eds.), Vol. 9. Metal Powder Industries Federation, Princeton, New Jersey.

Green, R. J. (1972). *Int. J. Mech. Sci.* **14**, 215–224.

Kaufman, S. M., and Mocarski, S. (1971). *Int. J. Powder Met.* **7**, 19–30.

Kuhn, H. A. (1972). *In* "Powder Metallurgy for High Performance Applications" (J. J. Burke and V. Weiss, eds.). Syracuse University Press, Syracuse, New York.

Kuhn, H. A., and Downey, C. L. (1971). *Int. J. Powder Met.* **7**, 15–21.

Kuhn, H. A., and Downey, C. L. (1973). *Trans. ASME* **95H**, 41–46.

Kuhn, H. A., Lee, P. W., and Erturk, T. (1973). *Trans. ASME* **95H**, 213–218.

Lee, P. W., and Kuhn, H. A. (1973). *Met. Trans.* **4**, 969–974.

Suh, S. K., and Kuhn, H. A. (1977). *In* "Modern Developments in Powder Metallurgy" (P. W. Taubenblat and H. H. Hausner, eds.), Vol. 9. Metal Powder Industries Federation, Princeton, New Jersey.

Suh, S. K., Kuhn, H. A., and Downey, C. L. (1976). *Trans. ASME* **98H**, 330–336.

Torre, C. (1948). *Hüttenmännische Monatsh.* **93**, 62.

Chapter 5

Analysis of Mechanical Property–Structure Relations in Powder Forgings

Alan Lawley

DEPARTMENT OF MATERIALS ENGINEERING
DREXEL UNIVERSITY
PHILADELPHIA, PENNSYLVANIA

I. INTRODUCTION

In the conventional powder metallurgy process, powder is compacted and then sintered to give parts of acceptable strength. Such parts are, in general, characterized by limited ductility, low toughness, and fatigue resistance. In consequence, their application is severely limited to nonload-bearing service environments. Residual porosity is the primary limitation on property levels; pores act as crack sites, centers of stress concentration and also aid in crack propagation. Repressing, infiltration, and post heat treatments serve to improve property levels. However, the magnitude of the improvement is insufficient to promote direct competition with cast and wrought materials in primarily load-bearing applications. As a guide, semidense parts with ~5% porosity are characterized by static strength levels $\leq 100,000$ psi (690 MPa) and a ductility $\leq 5\%$.

These limitations on mechanical property levels have served as a stimulus in the development of high-performance full-density powder metallurgy technology. Production of structural components from powder, on a competitive basis vis-à-vis wrought products, requires that the final material be free of porosity. Though several approaches can be followed in the quest for full density all fall in one of two major categories, namely:

(i)   direct densification of the powder; and
(ii) preparation of a porous preform followed by working of the preform.

Hot isostatic pressing is an example of the first route. Preform forging, rolling, or extrusion are deformation processing modes representative of the second route. In Chapter 4 it was pointed out that powder forging is now a viable commercial method of fabrication for high strength machine parts, while rolling and extrusion are at the laboratory or prototype stage of development.

The intent in this chapter is to evaluate and analyze mechanical behavior and property integrity of powder forged materials. Particular interest and importance are vested in dynamic properties such as toughness and fatigue resistance. It will be shown conclusively that while all porosity must be eliminated, this is not a sufficient criterion to guarantee optimal resistance to crack propagation in impact or cyclic loading. Rather, the mode of densification, i.e., repressing versus upset forging, is a further critical factor. The nomenclature and details of the powder forging process per se discussed in Chapter 4 serve as a useful precursor to the present mechanical property processing analysis.

II. DENSIFICATION AND MODE OF PORE CLOSURE

In subsequent sections of this chapter, mechanical behavior and property levels will be documented and analyzed in terms of the level of porosity and mode of pore closure occurring during densification in preform forging.

Thus, it is appropriate to reiterate and enlarge upon the concepts and fundamentals of densification discussed in Chapter 4.

First, there are essentially two distinct regimes of powder forging, namely, repressing and upset forging. In repressing, extensive flow of material in the lateral direction is prevented by die wall constraint. As full density is approached, the state of stress approximates a pure hydrostatic condition. Upset forging is characterized by extensive unconstrained (free) lateral flow of material during forging. Under these conditions, deviatoric and hydro-static components of stress are present. Clearly, if during a later state of up-set forging the material reaches the die wall, then subsequent densification will be in the repressing mode. Second, from the viewpoint of the metallur-gist, the two regimes of powder forging are expected to give rise to differing levels of microstructural integrity.

A. Repressing

In repressing, there is a reduction in height of the preform but no lateral flow other than that required to take up the initial small clearance between the preform and the die wall. A typical pore is simply flattened, and the opposite sides of the pore are brought together under pressure. Thus, a spherical pore becomes ellipsoidal upon collapse with the long axis equal to that of the original pore diameter (Fig. 1(a)). In the absence of any shearing action, oxides or other contaminant layers on the internal surfaces of the pores can mitigate against the development of a strong mechanical bond across the interface.

Chemical bonding can occur as a result of atomic diffusion across the

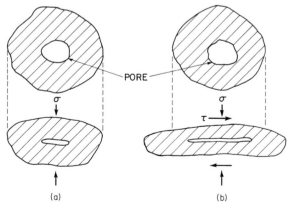

Fig. 1 Schematic representation of the mode of pore closure in (a) repressing and (b) upset forging.

collapsed pore interfaces. Clearly, process parameters that affect the diffusion rate will influence final bond integrity. From a synthesis of theory and experiment (Antes, 1974; Antes and Stockl, 1974; Cook, 1974, 1975; Ladanyi *et al.*, 1975) it has been shown that the following are of primary importance in the repressing of ferrous preforms: preform sintering temperature; repressing force; forging temperature; and cooling rate. Higher sintering temperatures ($\sim 2300°$F ($1483°$K)) result in a lower total oxide content in the porous preform. This aids effective bonding by diffusion across collapsed pores. Similarly, a more intimate pore surface contact results as the repressing load is increased. A high repressing temperature followed by slow cooling will increase the rate and extent of atomic diffusion across collapsed pore interfaces.

B. Upset Forging

True upset forging involves compressive height strain and lateral flow of the preform material. A spherical pore undergoes flattening and simultaneous elongation in the direction of lateral flow (Fig. 1(b)). Shear stress is now present along with the normal pressure and the former insures relative motion between opposite sides of the collapsed pore. Thus, mechanical rupturing of the oxide film is possible so that virgin material is exposed. This, along with intimate surface contact due to sliding, is expected to result in a sound metallurgical bond across collapsed pore interfaces. The flow of material during forging and densification will also produce fibering of inclusions in the lateral direction. Together, the inclusion morphology and that of collapsed pores introduce anisotropy in microstructure and hence in mechanical behavior.

C. Anisotropy and Resistance to Crack Propagation

Powder forging in the repressing mode does not lead to marked anisotropy. There is little or no effect on inclusion orientation, but some limited alignment of collapsed pores, nominally perpendicular to the direction of the forging load, will occur. Initial preform density dictates the extent of collapsed pore anisotropy, since even in repressing, material flow on a local scale must accompany densification. A low preform density will result in more local flow and hence a higher degree of anisotropy than in a preform of high initial density (Moyer, 1974). These effects not withstanding, in a repress powder forging, cracks propagate through a relatively isotropic structure. Impact resistance should then be independent of the orientation of the test piece cut from the powder forging.

In comparison, upset forging produces anisotropy of both collapsed pores and inclusions; the latter will have been converted into stringers as a result of lateral flow of the preform during densification. Consider now the propagation of a crack through this microstructure. In the longitudinal orientation, the crack front must propagate perpendicular to the direction of material flow. This direction is also perpendicular to planes of weakness in the form of collapsed pores and strung-out inclusions. For the short-transverse orientation, the direction of propagation of the crack front will be parallel to the direction of flow and also parallel to the planes of weakness. Thus, in upset forgings, it is expected that resistance to crack propagation and, hence, toughness will be higher in the longitudinal orientation than in the short-transverse orientation. Such behavior has been observed in hot-rolled plate; the anisotropy in this case is due solely to mechanical fibering of inclusions (Dieter, 1976).

The various flaws in the preform (e.g., residual pores, collapsed pores, and inclusions) give rise to local elastic (and possible plastic) strains. Morphology of the flaws is important since the magnitude of local internal stress concentrations is dependent on flaw size, shape, and the mode of external loading. A complete understanding of crack propagation and fracture necessitates analysis of the state of stress. For coherency, sample analyses are included along with a consideration of mechanical property levels in subsequent sections of the chapter.

We first examine the effect of porosity on property integrity in powder forgings of less than full density. This is followed by a consideration of the mechanical properties of fully dense powder forgings, with particular reference to the mode of densification, i.e., repress versus upset forging.

III. POROSITY AND MECHANICAL BEHAVIOR

In powder materials of less-than-full density, the level of porosity is the major variable controlling mechanical properties. The effect of porosity per se is best evaluated by considering mechanical property levels as a function of density in repress forgings. This essentially eliminates material flow as a contributing factor. When mechanical property data pertain to upset forging, judgment is necessary in interpretation because of the accompanying material flow.

A. Static Properties

Yield strength, tensile strength, ductility, and elastic modulus each decline as the amount of porosity increases. The general effect is shown schematically in Fig. 2 for tensile strength and elongation to fracture.

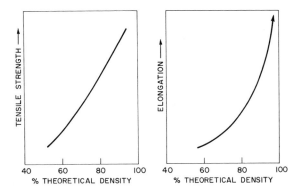

Fig. 2 The dependence of (a) tensile strength and (b) ductility on density in powder materials—schematic.

Kaufman and Mocarski (1971) have shown conclusively that the yield, tensile strength, and reduction in area of low-alloy powder forgings increase with decreasing porosity. Densities in the range 95–>99% theoretical were obtained by hot closed-die forging (repressing). Properties increase linearly with density to ~98% of theoretical; at higher density levels, property levels increase at a higher rate confirming the strong influence of the last traces of porosity.

A similar response has been observed for mixed elemental nickel steels (Chung and Koczak, 1977). In this case, the powder preforms were cold repressed to densities in the range 91–97% of theoretical. Initial preform density was also shown to influence strength. At a given repress density, tensile strength is higher for preforms of lower initial density. This is consistent with the greater amounts of local material flow characteristic of the repressing operation in a low-density preform.

The effect of porosity on strength during densification by preform working can be analyzed using a simple model. Consider an ideal porous material of relative strength σ_r as defined by Haynes (1970):

$$\sigma_r = \sigma/\sigma_0 = (1 - \rho) \tag{1}$$

where σ is the true stress for flow of the porous preform at a specified level of strain, σ_0 the true stress for flow of the fully dense material at the same level of strain, and ρ the percent porosity. In a real (nonideal) situation, the pores give rise to local stress concentrations in addition to reducing the effective load-bearing cross section. If the stress concentration factor due to pores is K_p, Eq. (1) is modified into

$$\sigma_r = \sigma/\sigma_0 = (1 - \rho)/K_p \tag{2}$$

In compressive upsetting of a porous preform, as densification occurs, p is decreased and both σ_r and K_p approach unity.

This type of behavior has been confirmed experimentally for the compressive upsetting (homogeneous deformation) of sintered iron preforms (Koczak and Lawley, 1972). Figure 3 shows the reduction in stress concentration with increasing density for a sponge iron powder. By examining various particle size fractions of a spherical iron powder, it was also shown that K_p is higher for large pores than for small pores at a given forged density level.

B. Dynamic Response

1. Toughness

The toughness of a forged powder preform is extremely sensitive to porosity; a typical impact resistance versus densification curve is depicted in Fig. 4.

Ishimaru et al. (1971) have obtained a fourfold increase in the impact resistance of low carbon steel powder forgings over the density range 92–~99% of theoretical. Since densification involved closed-die forging of the preforms, the enhancement in toughness can be attributed directly to a reduction in pore content. Similar increases in impact energy have been noted in mixed elemental nickel steel powder preforms cold repressed to densities in the range 91–~95% of theoretical (Chung and Koczak, 1977).

As an example of nonferrous compositions, the toughness of aluminum alloy 601 AB powder forgings has been shown to be improved by densifica-

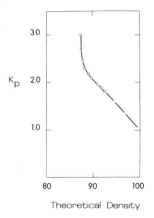

Fig. 3 Dependence of the stress concentration factor K_p on density for upset-forged sponge iron preforms. (Koczak and Lawley, 1972.)

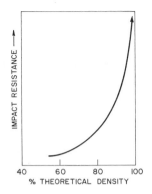

Fig. 4 The variation of toughness (measured in terms of impact resistance) with density—schematic.

tion (Ferguson and Lawley, 1974). Impact data obtained on subsize Izod test pieces are illustrated in Fig. 5. In this study, preforms in the shape of right cylinders were hot upset under conditions of homogeneous deformation. As explained by Kuhn, Chapter 4, the absence of friction between preform and die faces coupled with free lateral flow of material does not lead to full density at realistic height strains. Note the rapid increase in toughness at density levels $\gtrsim 98\%$ of theoretical. The levels of preform height strain (and hence extent of lateral flow) required to achieve densities $\gtrsim 98\%$ of theoretical were similar, e.g., in the range 75.5–78.9%. Hence it can be concluded that the remaining porosity is a major factor vis-à-vis toughness.

The study by Ferguson and Lawley (1974) also demonstrated the existence and importance of anisotropy. Impact data on subsize Izod test pieces notched in different orientations relative to the direction of forging are shown in Fig. 6. It is seen that, at a given density, specimens notched perpendicular to the forging direction exhibit higher levels of toughness than specimens notched parallel to the forging direction.

Anisotropy is the result of texture arising from the deformed pore structure and from orientation of flow lines in the aluminum alloy powder forgings. Macroscopically, the overall direction of crack propagation in impact test pieces notched parallel to the forging direction is along the flow lines (Fig. 7(a)). In comparison, propagation is across the direction of material flow when the test piece is notched perpendicular to the forging direction (Fig. 7(b)). The orientation of pores, relative to the test piece notch, is also illustrated in Fig. 7; pores flatten in planes approximately perpendicular to the forging direction. The pore morphology in Fig. 7(b) is more effective in arresting a moving crack than is the other morphology (Fig. 7(a)). There is more material behind the notch in advance of the crack front in the former

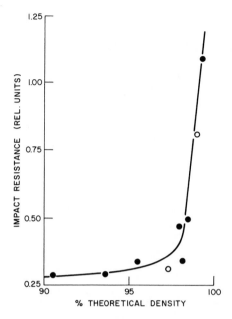

Fig. 5   Impact resistance of hot upset 601 AB aluminum alloy powder preforms as a function of density. (●) Homogeneous deformation (teflon lubricant); (○) are for forging without lubricant. (Ferguson and Lawley, 1974.)

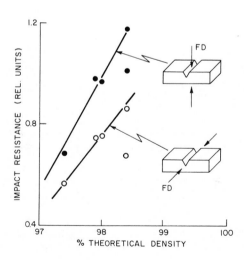

Fig. 6   Anisotropy of impact resistance in hot upset 601 AB aluminum alloy powder preforms as a function of density. All data points are for homogeneous deformation (teflon lubricant); FD is forging direction. (Ferguson and Lawley, 1974.)

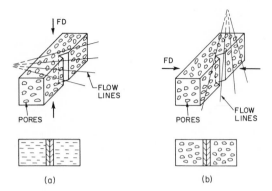

Fig. 7 A schematic of flow line orientation and collapsed pore morphology/anisotropy in Izod test pieces cut from hot upset 601 AB aluminum alloy powder preforms. (a) Notched parallel and (b) perpendicular to the forging direction (FD). (Ferguson and Lawley, 1974.)

(Fig. 7(b)) than in the latter (Fig. 7(a)). A comparison of the two corresponding fracture surfaces is made in Fig. 8 at the same final forging density (98.3% theoretical) and initial preform density (80% theoretical). Both test pieces were actually cut from the same forging. Qualitatively, the fracture surface in Fig. 8(a) is significantly rougher than in Fig. 8(b), reflecting the larger amount of energy required to propagate the crack in the former.

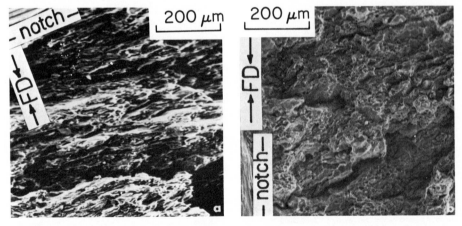

Fig. 8 Scanning electron micrographs of impact test piece fracture surfaces of hot upset 601 AB aluminum alloy powder preforms, (a) Notched perpendicular and (b) notched parallel to the forging direction. (Ferguson and Lawley, 1974.)

2. Fatigue Resistance

Systematic studies of the effect of porosity on the fatigue response of powder forgings have not been made. Data that are available pertain to forged preforms approximating full density, i.e., >98% theoretical. To evaluate the role of porosity under conditions of cyclic loading, it is necessary to resort to observed property levels for sintered compacts. Typically, these materials have densities in the range ~80–92% of theoretical.

As might be expected, the fatigue strength of sintered powder compacts increases as residual porosity decreases (Kravic and Pasquine, 1969; Bankowski and Feilbach, 1970; Haynes, 1970; Williams and Haynes, 1973). The effect is shown schematically in Fig. 9. To cite a specific example, the endurance limit of as-sintered nickel steels (Kravic and Pasquine, 1969) is increased from 19,000 psi (131 MPa) to 30,500 psi (210 MPa) as the sintered density is raised from 84 to 92% theoretical. In the quenched and tempered condition, the corresponding endurance limits over this range of densities are 35,000 psi (241 MPa) and 45,000 psi (310 MPa), respectively.

In ferrous compositions, the *S–N* curves exhibit a sharp knee similar to that in cast and wrought steels. The endurance ratio (fatigue limit:tensile strength ratio) is typically ~0.4. Since this ratio increases as density increases, this means that as the last traces of porosity are removed, fatigue strength increases more rapidly than tensile strength. Thus, optimization of fatigue performance demands that all porosity be removed.

Generally, fatigue cracks in sintered powder compacts initiate at free surfaces. Propagation of cracks is then aided by the linking up of pores.

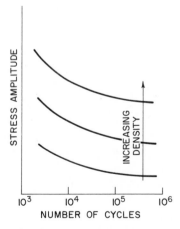

Fig. 9 The dependence of response to cyclic loading on density in powder materials— schematic.

While porosity thus aids propagation, it also is effective in blunting cracks so that failure is not catastrophic. It is interesting to note that because of these internal stress raisers, sintered semidense powder materials are less affected by external notches than fully dense cast and wrought stock. As porosity is removed, internal stress gradients are reduced and fatigue resistance increases toward levels characteristic of the conventional cast or wrought condition.

The content and morphology of inclusions also influences fatigue resistance. These flaws act as internal crack initiation sites, crack arrestors, or as links for the joining up of cracks. According to Brown (1974), with increasing density a transition point is reached beyond which inclusions probably become more dominant in controlling fatigue strength and crack propagation resistance than residual porosity.

IV. MECHANICAL BEHAVIOR AT FULL DENSITY

Thus far, we have characterized and established a rationale for the role of porosity and inclusions on strength integrity and crack propagation resistance in powder forgings and sintered material. It is now pertinent to consider the mechanical behavior of fully dense materials densified by the forging of a powder preform. In particular, we will be evaluating and comparing the properties of repress and upset-forged preforms. Experimental results and observations will be analyzed in light of the associated mode(s) of pore closure and any accompanying anisotropy of pore structure, inclusions or other flaws.

A. Static Properties

The powder metallurgy literature is now replete with information and data on the tensile properties of fully dense powder forgings. Materials evaluated include iron, a wide range of ferrous compositions, aluminum alloys, titanium, and cobalt or nickel-base superalloys. It is recommended that the reader refer to selected conference proceedings in order to gain familiarity with this literature base (Burke and Weiss, 1972; Hausner and Smith, 1974; Taubenblatt and Hausner, 1977).

It can be concluded from these studies that with few exceptions, tensile strength is comparable in repressed and upset-forged preforms at full density. A similar conclusion can be drawn regarding ductility. Furthermore, the levels of strength and ductility achieved in powder forgings are equal to (or can be higher than) those of corresponding cast and wrought compositions. This implies that under the loading environment of the standard tesile test, bond integrity of collapsed pores can be achieved without the need for shear and lateral flow of material during densification.

If die wear were the only consideration, repressing would be favored over upset forging. In upset forging, significant relative motion occurs between die surfaces and the powder preform during densification. It is also posssible that some die wall erosion can occur as the material moving laterally outwards impacts the die sidewalls. Thus, from the standpoint of die wear, repressing is more attractive than upset forging. This factor cannot be considered in isolation however, since it has been shown previously (Section II) that high temperature and forging loads are frequently necessary to insure interface pore bond integrity in repressing. In practice it may not be feasible or economical to operate under these conditions.

Die chill is a further factor that influences the choice of forging mode, particularly with ferrous powders. In the repressing of a porous preform, die chill from sidewall contact lowers preform temperature, resulting in an increase in yield strength and, in turn, in forging load. In comparison, the die chill problem is much less severe in upset forging. Though the material contacts the die sidewall after lateral flow, i.e., at the later stages of upset forging, densification is nearly complete by this time; hence any effect of a reduced temperature on load increase will be minimal. Because of the die chill effect and the above considerations of forging load needed to insure collapsed pore bond integrity, it is clear that higher forging loads are necessary in repressing than upset forging in order to achieve full density.

B. Dynamic Response

1. Toughness

In considering toughness, the major objective here will be to relate this dynamic property to forging mode and in turn to microstructural integrity at full density.

Various amounts of lateral flow and shear are possible in powder forging by upsetting preforms of equal volume but differing heights. Larger heights provide greater clearance between the preform and die walls and, therefore, result in greater lateral flow before any densification in the repressing mode occurs. The effectiveness of the various amounts of deformation in developing a sound bond can then be assessed through Izod or Charpy impact tests; both tests provide a sensitive measure of the integrity of the metallurgical structure.

Representative Izod values for fully dense material cut from cylindrical powder forgings of 601 AB aluminium alloy in the F temper are illustrated in Fig. 10 (Kuhn and Downey, 1974). The subsize Izod test pieces were notched in the longitudinal orientation (i.e., notched perpendicular to the direction

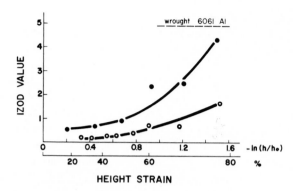

Fig. 10 Impact resistance of fully dense hot-forged 601 AB aluminum alloy powder preforms as a function of extent of height strain (and hence lateral flow). Preform density: ● 92%, ○ 82%. (Kuhn and Downey, 1974.)

of forging) so that the crack front propagated in the direction of forging. For purposes of comparison, data are also shown for wrought material of the same composition and using identical size and geometry for the Izod test piece. It is clear that, although all data points pertain to full density material, toughness is enhanced significantly by lateral flow in upset forging (high preform height strains). Note that impact resistance approached that of the wrought 6061 aluminium alloy for upset forgings of the higher starting preform density. Similar behavior is exhibited by a powder-forged low-alloy steel (4620) in the normalized condition at full density (Fig. 11). Here, a saturation (shelf) value is achieved after about 50% height strain.

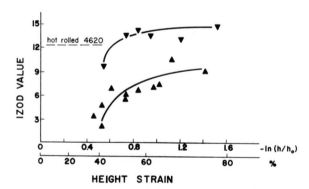

Fig. 11 Impact resistance of fully dense hot-forged 4620 steel powder preforms as a function of extent of height strain (and hence lateral flow). Preform density: ▼ 75%, ▲ 83%. (Kuhn and Downey, 1974.)

Impact fracture surfaces from the repressed and upset forgings reveal distinct differences (Koczak *et al.*, 1974). For aluminum, the repressed impact fracture surface is characterized by brittle interparticle fracture. With increasing lateral flow, the fracture morphology is transgranular, characteristic of high ductility. The improvement in toughness in the upset-forged material is attributed to breakup of the oxide layers that surround each powder particle surface. In the low-alloy steels, scanning electron microscopy confirms microporosity along particle boundaries, which remain open in the repressed condition. With increasing levels of lateral flow, this source of porosity is eliminated and in consequence the impact resistance increases. The fracture surfaces of the repressed forgings show interparticle separation; this mode of crack propagation is not evident in forgings subjected to lateral flow.

The above observations on toughness in aluminium alloy and low alloy steel powder forgings provide an important comparison on the effect of initial preform density. If material flow were the only consideration, maximum toughness is to be expected at some intermediate level of initial preform density. The rationale is that at low initial preform densities, much of the material flow serves only to close up porosity without significant accompanying shear and interparticle movement. At too high a starting preform density, insufficient deformation is imparted to the preform to provide adequate interparticle displacements and integrity of bonding across collapsed pore interfaces. Superimposed on flow will be the effect(s) of phenomena intrinsic to each material. In the case of aluminum alloy powders, the tenacious oxide film around powder particles must be broken up. In preforms of low initial density there is a greater pore surface area per unit volume and, therefore, a larger oxide content than in preforms of higher initial density. In consequence, the preforms of 92% initial density exhibit higher levels of toughness than those of 82% initial density after upset forgings to full density (Fig. 10.) By comparison, it is the lower initial preform density (75 cf 83% theoretical) that results in optimum toughness in low alloy steel powder forgings (Fig. 11). This is attributed to the fact that the lower density preforms have a significantly larger amount of interconnected porosity; during sintering, the oxide reducing gas can penetrate the lower density preforms to a greater extent than in high-density preforms. In this way a lower final oxide content exists prior to forging and this is instrumental in raising toughness (Kuhn and Downey, 1974). In other studies on iron-base powders, Antes (1972), Durdaller (1972), and Moyer (1972) have demonstrated a maximum in toughness after upset forging to full density for an initial preform density $\sim$78% theoretical.

A further confirmation of the enhancement of toughness by lateral flow in fully dense ferrous powder forgings is afforded by the observations of

Ferguson *et al.* (1975). 1000 and 4620 alloy compositions were hot-forged to full density under plane-strain conditions. Specific levels of lateral flow were achieved by controlling preform geometry (height:length); the initial preform density was 80% of theoretical in all cases. Charpy data for standard size test pieces of the as-forged 1000 iron powder (0.03% C, 0.3% Mn, 0.3% Cu) in the longitudinal orientation are shown in Fig. 12. There is a two-fold increase in impact resistance as the compressive height strain of the preform is increased from 30% (true strain—0.35) in repressing to ∼70% (true strain—1.2) for maximum lateral flow. The arrowed data points indicate incomplete fracture so that these are lower bound levels on impact resistance.

Similar increases in toughness with lateral flow, as measured on subsize Izod test pieces, occur in the as-forged (normalized) and heat-treated (tempered martensite) 4620 (0.2% C, 0.35% Mn, 0.5% Mo, 1.8% Ni) alloy steel powder forgings (Figs. 13 and 14). Corresponding impact levels for 4620 forged bar stock are also included for comparison in the figures. Apart from the role of lateral flow, the observations on the 4620 steels provide other insights into response to dynamic loading.

In the as-forged condition, impact resistance in the longitudinal orientation powder forgings approaches that of the bar stock tested in the same orientation (Fig. 13). After heat treatment, however, the toughness of the powder forgings is well below that of the bar stock (Fig. 14). This reflects a difference in strength level in the heat-treated powder and bar stock forgings; for example, 160,000 psi (1103 MPa) and 110,000 psi (758

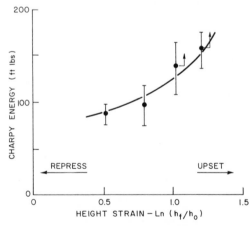

Fig. 12 Charpy impact response (standard test piece) for fully dense as-forged 1000 iron powder preforms as a function of extent of height strain (and hence lateral flow); 1 ft lb force = 1.36 J. (Ferguson *et al.*, 1975.)

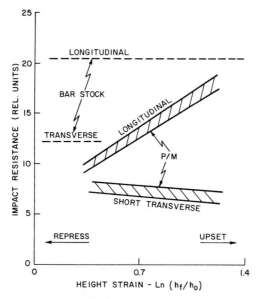

Fig. 13 Izod impact response of fully dense as-forged 4620 steel powder preforms as a function of extent of height strain (and hence lateral flow). (Ferguson *et al.*, 1975.)

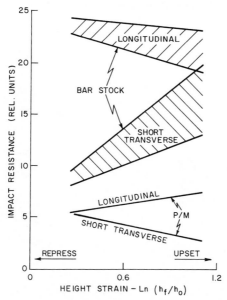

Fig. 14 Izod impact response of fully dense heat-treated 4620 steel powder forged preforms as a function of extent of height strain (and hence lateral flow). (Ferguson *et al.*, 1975.)

MPa), respectively, at maximum lateral flow. Since toughness and strength are inversely related (Tetelman and McEvily, 1967), the powder forgings will have a lower level of toughness than the forged bar stock. These differences in strength in the two materials preclude a direct comparison of impact resistance. Clearly, this does not detract from the benefits of lateral flow on toughness at a given strength level.

The 4620 powder forgings are anisotropic with respect to impact resistance in both the as-forged and heat-treated conditions. Further, the difference between longitudinal and short transverse toughness increases with increasing levels of lateral flow during forging (Figs. 13 and 14).

The earlier considerations of repressing and upsetting vis-à-vis mode of pore closure and inclusion morphology have served to establish an acceptable rationale for understanding and interpreting toughness integrity. A more complete analysis is provided by means of a detailed characterization of microstructure and fracture morphology. We will refer to the work of Ferguson (1976) and Ferguson *et al.* (1975) to illustrate microstructural analysis and interpretation.

Charpy fracture surfaces of repress and upset forged 1000 powders (at low magnification) are shown in Fig. 15. In the repress condition, the fracture surface is relatively flat, has an overall (macroscale) orientation perpendicular to the long axis of the test piece, and is accompanied by the formation of shear lips. In contrast, a slant fracture is evident in the true upset forged material. Here the crack propagating overall in the forging direction is deflected locally along flow lines, as illustrated in Fig. 16. Similar

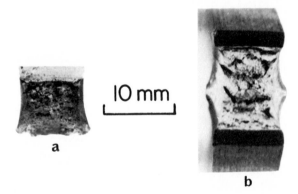

Fig. 15 Charpy impact fracture surfaces for 1000 iron powder forgings; (a) Repressed (30% height strain) and (b) upset forged (70% height strain) conditions. Longitudinal orientation test pieces.

changes in fracture mode and resulting fracture surface morphology with increasing amounts of lateral flow have been confirmed in the as-forged 4620 material. An example of the slant fracture profile in the upset forged condition is given in Fig. 17.

It is clear that, for material subjected to extensive lateral flow during forging, cracks propagating in the forging direction are blunted and deflected along flow lines. The mechanism responsible for this deflection and, in turn, enhanced longitudinal orientation impact resistance, is delamination. This occurs along bonded pore interfaces (by the reopening of collapsed porosity) or at matrix–inclusion interfaces. Anisotropy of collapsed pores and inclusions as a result of material flow is an essential ingredient; in repress forgings, anisotropy is minimal and crack front deflection does not occur. It is the energy associated with the process of delamination that gives rise to the increase in toughness.

Fig. 16 Charpy fracture surface profile in a fully dense upset-forged 1000 powder pre-form (height strain 70%). Longitudinal orientation test piece.

Fig. 17 Profile of the slant fracture surface in a fully dense upset-forged powder preform of 4620. Longitudinal orientation, as-forged after 70% height strain.

2. Fatigue Resistance

Several recent studies have confirmed that, like toughness, the fatigue strength and resistance to crack propagation under conditions of cyclic loading is enhanced by lateral flow at full density. The most complete experimental study and analysis, in terms of microstructure and stress state, is that of Ferguson (1976) and Ferguson *et al.* (1977). We will cite this study to illustrate the importance of a thorough characterization of macro- and microstructure in order to understand dynamic response.

In this study (Ferguson, 1976; Ferguson *et al.*, 1977) full-density 4620 steels were prepared by hot forging under plane strain conditions. High and low levels of lateral flow were achieved in true upset forging and repressing, respectively, through control of the preform geometry. The powder forgings were tested at room temperature in axial fatigue (at a stress ratio of 0.1) and in rotating bend fatigue. Both as-forged (normalized) and heat-treated (tempered martensite) structures were evaluated.

(a) Axial Fatigue Axial fatigue data are shown in the conventional form of *S–N* curves in Figs. 18 and 19 for the as-forged and heat-treated conditions, respectively. It is seen that the endurance limit is indeed raised as a result of lateral flow in both figures. Since fatigue strength is related to static strength, it is necessary to normalize the data and look at the variation of endurance ratio (fatigue limit: ultimate tensile strength) with lateral flow. This is done in Fig. 20; it is clear that in axial fatigue, material flow increases the fatigue resistance of the 4620 forgings in both the normalized and tempered martensite structures.

To establish the generality of the beneficial effect of flow in axial fatigue, a prototype powder metallurgy part was processed from 4620. The part was an axisymmetric multilevel flange–hub coupling, similar to a drive flange. Preform design for this part was considered in Chapter 4, Section

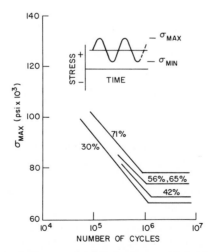

Fig. 18 Axial fatigue *S–N* curves for full-density powder forgings of 4620 as a function of height strain; as-forged condition. (1 ksi = 6.9 MPa.)

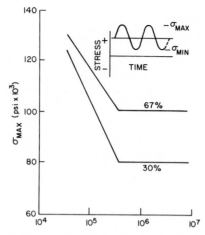

Fig. 19 Axial fatigue *S–N* curves for full-density powder forgings of 4620 as a function of height strain; heat-treated condition. (1 ksi = 6.9 MPa.)

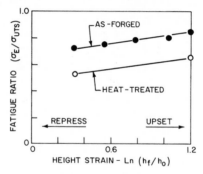

Fig. 20 The endurance ratio of 4620 powder forgings in axial fatigue as a function of height strain.

IIIC. By forging from two different preform geometries, two levels of flow were achieved. These parts were than tested in axial fatigue by supporting the part flange and applying load cyclically along the axis of the hub.

S–N curves for the prototype parts in the as-forged condition and for a part of identical size and geometry machined from 4620 bar stock at the same strength level are illustrated in Fig. 21. As in the conventional axial fatigue tests (Figs. 18 and 19), lateral flow accompanying upset forging is seen to enhance fatigue resistance. It is also clear that the parts machined from the bar stock are inferior to either conditions (repress or extruded) of the powder forgings.

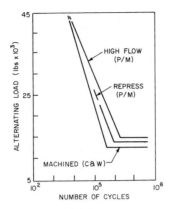

Fig. 21 *S–N* curves for simulated axial fatigue on prototype powder forged parts. C and W indicate cast and wrought. (1 lbf = 4.448 N.)

Microstructurally, the grain flow pattern in the two flow conditions is significantly different. Attention is drawn to the critical region between the hub wall and flange (Fig. 22). In the repress formed part, lateral flow of material is restricted to the immediate vicinity of the surface at the corner junction (Fig. 22(a)). In comparison, flow of material has occurred throughout the complete cross section of the part densified in the extrusion mode (Fig. 22(b)). These structural perturbations are reflected in differences in fracture mode and path. Thus, in the repress part, the primary fatigue crack propagates essentially perpendicular to the direction of material flow, i.e., normal to the direction of maximum tensile stress (Fig. 23). Once the crack enters the region of axial compaction, i.e., true repressing, it is deflected along collapsed pores and prior particle boundaries. Its direction of propagation is now roughly parallel to the flange base. Note the precise coincidence of the location of the change in crack propagation direction with the flow–repress boundary. In the extruded part (Fig. 24), fatigue cracks again initiate at the corner junction but then propagate completely through the part. Integrity of microstructure resulting from the material flow accompanying extrusion precludes planes of weakness so that there is no change in the direction of crack propagation. The influence of inclusions in causing the blunting, redirection, and delamination of a crack in the powder part is illustrated in Fig. 25. The lower fatigue limit of the heat-treated bar stock compared to the powder metallurgy (P/M) forged parts is a consequence of the unfavorable inclusion anisotropy. Inclusions originally elongated in the rolling direction deflect fatigue cracks in a direction parallel to the bore of the part (Fig. 26). This weakens the resistance of the flange–hub junction to cyclic stress.

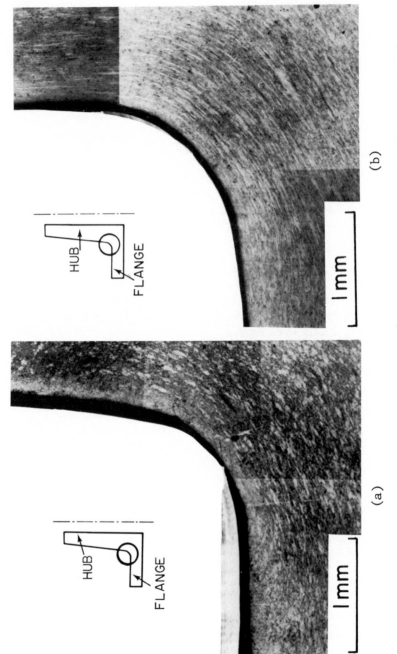

Fig. 22 Macroview of the grain flow pattern at the corner junction of a prototype powder part, (a) Repress and (b) extruded preform.

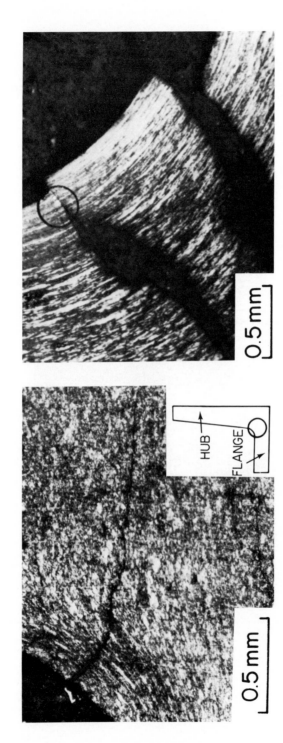

Fig. 24 Path of fatigue cracks in the extruded-hub prototype part.

Fig. 23 Path of primary fatigue crack in the repressed prototype part.

163

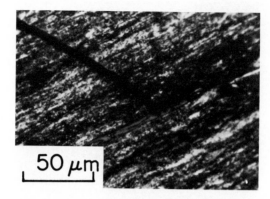

Fig. 25 Deflection of a fatigue crack by an elongated inclusion in the extruded-hub forging.

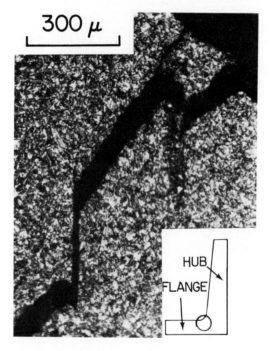

Fig. 26 Path of primary fatigue crack in the prototype part machined from bar stock.

(b) Rotating Bend Fatigue S–N curves for the rotating bend fatigue mode are shown in Figs. 27 and 28 for the as-forged and heat-treated conditions, respectively. In contrast to axial fatigue, the curves for repress and upset forgings superimpose. Furthermore, the endurance ratio is indepen-

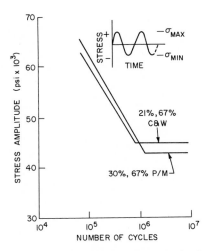

Fig. 27 Rotating bend *S–N* curves for full-density, powder forgings of 4620 as a function of height strain; as-forged condition. C and W is cast and wrought. (1 ksi = 6.9 MPa)

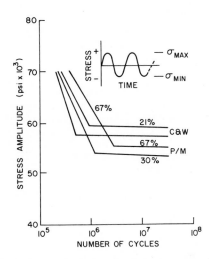

Fig. 28 Rotating bend *S–N* curves for full-density powder forgings of 4620 as a function of height strain; heat-treated condition. C and W is cast and wrought. (1 ksi = 6.9 MPa.)

dent of the extent of lateral flow (Fig. 29). Thus, in rotating bend fatigue, material flow does not significantly affect the fatigue resistance of the 4620 forgings in either the normalized or tempered martensite structures.

This difference in fatigue response has important implications in terms of

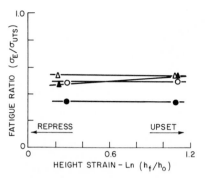

Fig. 29 The endurance ratio for 4620 powder forgings in rotating bend fatigue as a function of height strain. ○ as-forged P/M, ● heat-treated P/M, △ as-forged C and W, ▲ heat-treated C and W. C and W are for bar stock material.

in-service performance involving cyclic loading conditions. To rationalize the effect of fatigue anatomy on the role of lateral flow, it is necessary to consider the interaction of the cyclic stress state with the collapsed pores and inclusions in the fully dense preforms. In particular, we must examine low stress–high cycle ($> 10^4$) behavior since this established the lower limit of cyclic load level for a specific life span of the material before fatigue failure.

(c) Stress Analysis To arrive at some quantitative understanding of the levels of internal elastic stress concentration in a forged-powder preform, it is necessary to idealize the nature and geometry of flaws. Consider only voids and rigid (infinite elastic modulus) inclusions of circular or elliptical shape. In the case of the noncircular flaw, orientation relative to the direction of applied load will be a further variable.

(i) *Voids* For a void, the stress concentration is vested in the tangential stress at the void–matrix interface. Stress cannot be transmitted in the radial direction in the case of a void. The situation is depicted schematically in Fig. 30 for an elliptical void aligned with its long axis parallel to the direction of external loading. We consider both tension and compression.

In tensile loading, the tangential stress σ_θ is tensile and at a maximum at location A. At B, σ_θ is at a maximum compressive level. From elastic analysis (Dieter, 1976; D'Isa, 1968) the magnitude of the maximum tensile tangential stress (σ_θ max) at the void-matrix interface can be calculated from

$$\sigma_\theta(\text{max}) = \sigma_n(1 + 2a/b) \tag{3}$$

where σ_n is the nominal stress, a the ellipse axis perpendicular to the

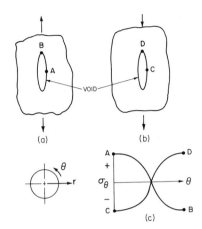

Fig. 30 Stress state around an elliptical hole for (a) tensile and (b) compressive loading.

applied load, and b the ellipse axis parallel to the applied load. Thus, in the case of a circular hole ($a = b$), σ_θ (max) is equal to $3\sigma_n$ at location A; i.e., the stress acts in the matrix at the ends of the hole in a direction parallel to σ_n. If the void is grossly elongated so that $b \gg a$, then σ_a(max) at A is reduced to a level approaching that of σ_n. The difference in magnitude between σ_θ(max) and σ_n is established by the degree of ellipticity of the void. The stress levels shown in Fig. 30(c) are simply illustrative of the variation of σ_θ with σ, not absolute values.

In the case of compressive loading, the situation is reversed. Now the maximum tangential stress at location C ($\equiv$ to location A for tension) is compressive. It is also seen that the tangential stress at D ($\equiv$ to location B in tension) is tensile and a maximum.

(ii) *Rigid inclusions* For a rigid flaw, stress can be transmitted in the radial direction across the matrix-inclusion interface, and hence it is necessary to consider radial (σ_r) as well as tangential (σ_θ) stresses. The stress levels as a function of θ are illustrated in Fig. 31.

Tensile loading produces a compressive radial stress at A but a tensile radial stress at B. The tangential component is negligible at A but is tensile at B. It is the radial stresses that are of primary concern since matrix–inclusion separation can occur if σ_r is tensile. Clearly, it is at location B that this separation will initiate under a normal tensile load. Elastic analysis shows that as b/a increases (i.e., for a highly elliptical inclusion with its long axis parallel to the loading direction), σ_r at B approaches σ_n, and the probability of crack initiation at the interface is decreased. It follows that rounded flaws are more prone to interface separation.

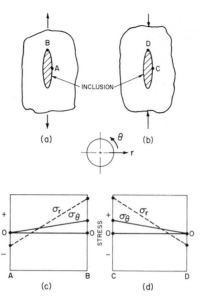

Fig. 31 Stress state around a noncircular rigid inclusion for tensile loading, (a) and (c), and compressive loading, (b) and (d).

If the external loading is *compressive*, the sense of the radial and tangential components of stress is reversed. Thus, inclusion–matrix separation is now likely to initiate at location *C*. An inclusion elongated in the direction of compressive loading is particularly suspect. Apart from debonding, the inclusion may fail in a buckling mode.

(d) Application of Analysis We are now in a position to consider the implication of the elasticity analysis in light of the observed fatigue response in axial or rotating bend modes. These two modes of testing must be considered for the two extremes in processing of the powder forgings, namely, repressing and upset forging.

In the axial fatigue studies carried out by Ferguson (1976) and Ferguson *et al.* (1977), the loading was tension–tension with a fatigue ratio ($\sigma_{min} : \sigma_{max}$) of 0.1. Thus, at all times during cyclic loading, the nominal load was tensile. In upset forged material the collapsed pores and inclusions are oriented in the direction of metal flow, which is also the direction of tensile loading. The analysis shows that this anisotropy of flaws minimizes stress concentration and hence crack initiation at collapsed pores or matrix–inclusion interfaces. Further, the collapsed pore morphology/anisotropy then deters crack propagation. Recall now the relative absence of marked anisotropy of collapsed pores or inclusions in repress forgings. Hence, both initiation and propagation of cracks can occur at a lower nominal stress as compared to the upset forged condition. This reflects in a lower level of fatigue strength and endurance limit in repress forgings than for upset forgings, when tested under a completely tensile cyclic load regime.

In rotating bend fatigue, both tension and compression are experienced during each revolution of the test piece. During the tensile part of the cycle, the situation is identical to that considered above in tension–tension fatigue. Hence the differences in fatigue strength manifest in the repress and upset forgings should be evident. We must, however, consider the situation during the compressive half cycle of load with respect to the upset-forged material. From the elastic analysis, the compressive stress gives rise to large stress raisers at collapsed pores and elongated inclusions in upset forgings.

In summary, the absence of flaw anisotropy in the repress material is detrimental to fatigue resistance in tensile loading. It is the presence of this anisotropy resulting from lateral flow in upset forging that is undesirable in compressive loading. Either way, in reverse stress-state fatigue, a similar response will be observed in the repress- and upset- forged powder preforms. In both conditions, the flaw morphologies preclude optimization of fatigue strength.

V.   SUMMARY AND CONCLUSIONS

In this chapter attention was focused on the mechanical behavior of powder forgings and on the correlation(s) that exist between mechanical properties and macro- or microstructure. While static (tensile) properties provide some measure of material quality, dynamic properties (toughness, fatigue strength) are much more sensitive monitors of structural integrity. Thus, provided full density is achieved, the strength and ductility levels of powder forgings will match (or exceed) those of conventional cast and wrought material. In contrast, full density is not a sufficient criterion by which it can be guaranteed that toughness or fatigue resistance of the powder forging will be as good as that of cast and wrought material. When these dynamic properties are under consideration, the route taken to achieve full density is a further critical factor. This is understood in terms of the mode of pore closure accompanying densification; two extreme conditions are identified, namely, repressing and upset forging. As a general rule, lateral flow of material in the upset forging mode enhances dynamic property levels by virtue of improved bonding across collapsed pore interfaces. The effect of lateral flow on inclusion morphology and anisotropy must also be considered in the evaluation of mechanical response. Analysis of the state of stress developed around residual pores, collapsed pores, and inclusions, for differing regimes of external loading, provides a basis for understanding fatigue response.

REFERENCES

Antes, H. W. (1972). *Progr. Powder Met.* **28**, 303.
Antes, H. W. (1974). *Metals Eng. Quart.* **14**, 8–15.
Antes, H. W., and Stockl, P. L. (1974). *Powder Met.* **17**, 178–192.
Bankowski, R. S., and Feilbach, W. H. (1970). *Int. J. Powder Met.* **6**, 23–37.
Brown, G. T. (1974). *Powder Met.* **17**, 103.
Burke, J. J., and Weiss, V., eds. (1972). "Powder Metallurgy for High Performance Applications." Syracuse University Press, Syracuse, New York.
Chung, H., and Koczak, M. J. (1977). *In* "Modern Developments in Powder Metallurgy" (P. W. Taubenblat and H. H. Hausner, eds.), Vol. 9. Metal Powder Industries Federation, Princeton, New Jersey.
Cook, J. P. (1974). *Int. J. Powder Met. Powder Tech.* **10**, 15–19.
Cook, J. P. (1975). *Metal Progr.* **107**, 85–88.
Dieter, G. E. (1976). "Mechanical Metallurgy," 2nd ed. McGraw-Hill, New York.
D'Isa, F. A. (1968). "Mechanics of Materials." addison-Wesley, Reading, Massachusetts.
Durdaller, C. (1972). *Metal Progr.* **101**, 44–49.
Ferguson, B. L. (1976). Toughness and Fatigue of Iron-Base P/M Forgings Ph.D. dissertation, Department of Materials Engineering, Drexel University, Philadelphia, Pennsylvania.
Ferguson, B. L., and Lawley, A. (1974). *In* "Modern Developments in Powder Metallurgy" (H. H. Hausner and W. E. Smith, eds.), Vol. 7, pp. 485–501. Metal Powder Industries Federation, Princeton, New Jersey.
Ferguson, B. L., Suh, S. K., and Lawley, A. (1975). *Int. J. Powder Met. Powder Tech.* **11**, 263–275.
Ferguson, B. L., Kuhn, H. A., and Lawley, A. (1977). *In* "Modern Developments in Powder Metallurgy" (P. M. Taubenblat and H. H. Hausner, eds.), Vol. 9. Metal Powder Industries Federation, Princeton, New Jersey.
Haynes, R. (1970). *Powder Met.* **13**, 465.
Hausner, H. H., and Smith, W. E. eds. (1974). "Medern Developments in Powder Metallurgy," Vols. 6, 7, and 8. Metal Powder Industries Federation, Princeton, New Jersey.
Ishimaru, Y., Yamaguchi, T., Saito, Y., and Nishino, Y. (1971). *Powder Met. Int.* **3**, 126–130.
Kaufman, S. M., and Mocarski, S. (1971). *Int. J. Powder Met.* **7**, 19–30.
Koczak, M. J., and Lawley, A. (1972). *Powder Met. Int.* **4**, 186–191.
Koczak, M. J., Downey, C. L., and Kuhn, H. A. (1974). *Powder Met. Int.* **6**, 13–16.
Kravic, A. F., and Pasquine, D. L. (1969). *Int. J. Powder Met.* **5**, 45.
Kuhn, H. A., and Downey, C. L. (1974). *Int. J. Powder Met.* **10**, 23.
Ladanyi, T. J., Meyers, G. A., Pilliar, R. M., and Weatherly, G. C. (1975). *Met. Trans* **6A**, 2037–2048.
Moyer, K. H. (1972). *Progr. Powder Met.* **28**, 5.
Moyer, K. H. (1974). *In* "Modern Developments in Powder Metallurgy" (H. H. Hausner and W. E. Smith, eds.), Vol. 7, pp. 235–254. Metal Powder Industries Federation, Princeton, New Jersey.

Taubenblat, P. W., and Hausner, H. H., eds. (1977). "Modern Developments in Powder Metallurgy," Vols. 9–11. Metal Powder Industries Federation, Princeton, New Jersey.

Tetelman, A. S., and McEvily, A. J., Jr. (1967). "Fracture of Structural Materials." Wiley, New York.

Williams, S. H., and Haynes, R. (1973). *Powder Met.* **16**, 387–403.

Chapter 6

Economic Risk Analysis of a Miniplant for Production of Steel Strip from Powder

Anthony R. Zecca

RESEARCH AND TECHNOLOGY
ARMCO STEEL CORPORATION
MIDDLETOWN, OHIO

George E. Dieter

PROCESSING RESEARCH INSTITUTE
CARNEGIE MELLON UNIVERSITY
PITTSBURGH, PENNSYLVANIA

Robert D. Thibodeau

SPECIAL PROJECTS DEPARTMENT
ATLANTIC RICHFIELD COMPANY
LOS ANGELES, CALIFORNIA

I. GENERAL INTRODUCTION

The decision to invest in a new steel plant is of major importance because the capital requirement is large and the financial effects long-lasting. In recent years, inflation, the economies of scale, and more stringent environmental legislation have sharply increased the investment needed for an integrated steel plant. As a result, nonintegrated minimills for producing light sections are growing in importance because of their low capital cost. In addition, they aid the environment by consuming scrap while their electric furnaces emit little pollution (Varga and Lownie, 1969; Regan et al., 1972). This chapter analyzes the extension of the minimill concept to sheet production by utilizing powder metallurgy (P/M) techniques. This idea was suggested decades ago but has yet to be adopted on a commercial scale due to technical and economic uncertainties.

The economies of steelmaking are uncertain because the prices of many factors of production, particularly scrap, have followed an erratic course. In such an environment, conventional business planning becomes unreliable. Risk analysis, originated by Hertz (1964), is a useful aid to decision making under these circumstances. The technique samples from the many possible scenarios to determine the probability of each outcome. Given the range and likelihood of the possible outcomes of P/M steel sheet production, it becomes possible to determine the desirability of the process.

II. RATIONALE AND PRINCIPLES OF RISK ANALYSIS

A. Capital Investment Decisions under Uncertainty

1. Risk and Uncertainty

Few capital investments in new processes or plants are undertaken with complete knowledge of the outcome. Several important project parameters such as product demand and price may be determined by other than the producer. Raw materials supply and government regulations pose additional problems. In a typical project, historical data provide a basis for estimating the value of some of the variables. Research, marketing, and other functions refine these estimates and also provide information relating to technological factors and potential actions of competitors. The net result of the planning process is that upper and lower limits of the value of each parameter are estimated. In addition, for each variable, a subjective estimate of the likelihood of each value is obtained. When this information is available, the decision to proceed with or abandon the project is strictly classified as decision making under risk (Ackoff and Sasieni, 1968). Risk analysis is one method of evaluating an investment under these circumstances and is the principal topic of this chapter.

Although incomplete knowledge is the usual background for investment decisions, it is useful to consider the extreme situations where knowledge is either complete or wholly lacking. In deterministic problems, the principles of capital budgeting or engineering economy can be used directly to arrive at the best decision. Lease-or-buy decisions or replacement decisions are typical cases of deterministic problems (Bierman and Smidt, 1969). At the opposite end of the spectrum are decision problems in which either the outcomes are unknown or the likelihood of possible outcomes cannot be estimated. This area is properly termed decision making under uncertainty. Marketing decisions involving competition fall into this class. A variety of decision criteria have been proposed for these problems and are reviewed by Luce and Raiffa (1957).

Both deterministic and uncertain investment problems are limiting cases of decision making under risk. When the uncertainties concerning a capital investment are minimal, a unique decision is reached provided an interest rate is assumed. As uncertainty increases, the correct decision is also a function of the value system of the decision maker. Thus, many investment opportunities are commonly accepted by one individual or organization although deemed too risky by others. The rationalization of this type of behavior is usually accomplished with the theory of utility. Savage (1972) explores the mathematical development of the theory while Schlaifer (1969) provides a useful handbook for the pragmatic business decision maker. That even utility theory is subjective has been noted by Mitroff and Betz (1972) so that there may, in fact, be no "right" investment decision at all. This need not be of concern at the moment for we will assume that utilities are equivalent to cash flow.

2. Advantages of Decentralized Decisions

In a complex investment such as a P/M steel plant, it is obvious that there are many sources of risk or indeterminacy. Some are outside of the plant boundaries in the overall socioeconomic environment. The price and availability of each raw material and the price and competitive situation of steel sheet are usually determined by a complex interaction outside the control of the plant management. Inside the plant are other uncertainties which affect profitability, including worker productivity, process yield, and machine breakdown. Some factors such as wage rates fall into both categories. In any case, there exist a great number of uncertain variables which, acting simultaneously, produce an uncertain economic outcome.

In conventional investment decision making, the emphasis is on reaching a consensus on what outcome will occur from an investment. Once this is agreed upon, an attempt is usually made by a high level decision maker to evaluate the overall risk of the project. This is invariably an impossibly complex task so that often one of the following two courses of action is chosen. In the first, usually called a sensitivity analysis, the effects of variations in each factor are investigated one at a time. Unfortunately real factors act in concert to produce unintuitive results. After examining the many sensitivity analyses, the decision maker often proceeds along the second path. By this time he is certain of only one thing—that an enormous number of outcomes are possible for the investment. He then estimates the worst outcome that could reasonably occur and makes his decision accordingly. In short, the decision maker has maximized his minimum gain. The shortcoming of this approach is that the effect of all optimistic outcomes is ignored.

The source of the dilemma is that one cannot consider simultaneous variations of many factors. One advantage of risk analysis is that the uncertainty of each factor may be estimated independently. Thus, in risk analysis, the marketing department estimates the probability of each level of demand. At the same time, purchasing is pondering over potential raw material prices. In this way each department determines the risk in its own area of expertise only. The result is a number of probability distributions relating the state of knowledge about each factor. If these distributions, instead of point estimates, are used as input for a standard economic analysis, the overall risk of the project becomes explicit. The result is a probability distribution of the return on investment. Sensitivity analyses are necessary only to show the relative importance of each individual factor. In addition, each uncertainty has been estimated by those who should be most qualified.

B. Engineering Model Construction Principles

The first step of an economic analysis of a process is to construct a material flow diagram based on the conservation of mass. A basis is selected and all other quantities are expressed in relation to that basis. Figure 1 shows the material flow diagram for the P/M steel plant. The basis is 175,000 tons

(157,000,000 kg) of hot-rolled and cold-rolled strip, sheet, and coil per year. The numbers in Fig. 1 indicate the flow of steel to various processes, in thousands of tons per year. At most stages there is a yield loss which is shown below each process as scrap or waste. Note that the scrap generated is recharged into the electric furnace as plant scrap.

The material flow diagram is extremely important because it allows visualization of the process. Another benefit is that each process unit can be scaled properly in relation to other units independent of the numerical value of the basis. For the P/M steel plant only the flow of steel is shown, although for more complex processes a mass balance for each element and an energy balance may be necessary. Williams and Johnson (1958) and Schuhmann (1952) provide additional details on material and energy balances.

C. Financial Models

1. Cash Flow Analysis

The use of discounted cash flow techniques permits a more realistic evaluation of a project than a simplified analysis which does not consider the timing of receipts and disbursements. Detailed descriptions of discounted cash flow analysis may be found in textbooks such as Bierman and Smidt (1969), or in the journal *The Engineering Economist*.

It is first necessary to define cash flow to distinguish it from profit. Gross profit is defined by accountants as the difference between revenue and costs. Included in the cost terms is depreciation, an allowance for the gradual deterioration of an asset. However, it should be noted that depreciation requires no cash outlay but is merely a bookkeeping entry. Income taxes are based on gross profit so that net profit is gross profit less taxes. The actual net amount of cash received by the company from operations is thus equal to the net profit plus depreciation. This is called the operating cash flow. The company may choose to make additional investments in working capital or in new equipment. These require an actual cash outlay and so are deducted from the operating cash flow to obtain the net cash flow. Simply stated, the net cash flow is the difference between cash receipts and cash outlays.

Cash flows rather than profits are used in investment analysis because they are precisely defined in terms of acts and events. The time of a particular cash flow is determined by the moment cash is actually spent or received. On the other hand, profit is a nebulous quantity strongly affected by variations in accounting practices.

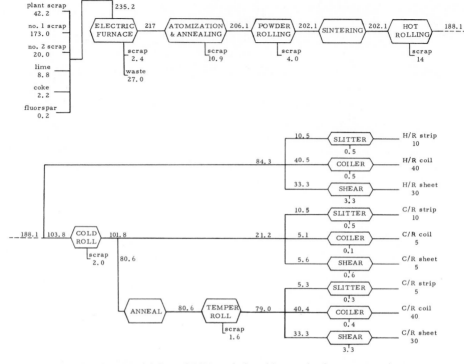

Fig. 1 Material flow of P/M steel plant (thousands of tons per year).

2. Effects of Time

It is obvious that mere knowledge of the cash flow of a project is not enough to determine its desirability. Timing of the cash flow is also of the utmost importance because idle cash can be invested at a profit.

Unfortunately, disbursements and receipts occur at different times so that a method is needed to place them on a common basis. During the initial years of a project, cash flows out of the company to pay for the capital investment. After the plant begins operations, cash begins to flow back into the company. One way of adjusting for the timing is to convert all future cash flows into current cash equivalents based on an assumed interest rate. This interest rate represents the earnings possible from alternative investments. For conversion

$$P = S/(1 + i)^N \tag{1}$$

where P is the present value, S the future value, i the interest rate, and N the number of years elapsed. The present worth factor is $(1 + i)^{-N}$. In most cases, however, it is more convenient to utilize tables of the present worth factor, which are available for different interest rates and time periods.

Equation (1) indicates that both higher interest rates and longer elapsed times decrease the present worth of a dollar. Because most projects require an immediate capital expenditure and receipts occur only after several years, it is apparent from Eq. (1) that higher interest rates decrease the present worth of an investment. Ultimately an interest rate is reached which makes the present value equal to zero. This rate is called the discounted rate of return on investment. It is equivalent to the maximum interest rate on a loan equal to the amount of the original investment which could be paid back out of the proceeds of the project.

3. Decision Rules

Two different statements of the decision rule for capital investments are shown by Bierman and Smidt (1969) to be equivalent under most circumstances. The first statement is to accept all projects having a positive present worth and reject all others. It should be noted that an interest rate must be assumed to do the present worth calculation. Another possibility is to determine the discounted rate of return on investment by trial-and-error and to accept all projects having a higher rate than some specified minimum rate of return. The use of either criterion requires that an interest rate representing alternative uses of capital be assumed.

D. Sources of Uncertainty

Investment decision making is much more than the application of discounted cash flow analysis techniques. If it were that straightforward, the duty could be delegated to anyone who could operate a desk calculator. The reason that organizations have traditionally reserved these decisions for top managers is that the cash flows for a project are always uncertain. Even a simple loan transaction with a fixed repayment schedule has a potential for default. In this section a number of important areas of uncertainty will be identified.

1. Exogenous Factors

Usually the most important sources of uncertainty in planning capital investment are the demand for and the price of the finished product. Raw materials prices have a smaller effect. According to AISI (1976), net earnings for the steel industry have been about 5% of sales revenue, so that small price changes have the potential for causing large variations in profits. An additional complexity is that price and demand are not independent but are correlated by a varying elasticity of demand, as shown by Mo and Wang (1970). A more subtle effect is that changes in demand cause highly nonlinear variations in production costs and, thus, in profits. The level of overall economic activity also exerts a strong influence on the demand for steel and cannot be controlled or even easily predicted (Klein, 1975).

The scaling of a steel plant is impacted by both internal and external factors. The most obvious and important external factors are the relationships between the size of the proposed plant, the size of the markets to be served, and the actions of competitors. There is no agreement on an optimum size; Leckie and Morris (1968) conclude that larger is better, while Cartwright (1972) emphasizes the advantages of smaller plants.

The cost of capital for an investment is, by definition, the weighted average of debt and equity capital for a particular company. Although estimation of the cost of debt should present few difficulties, the cost of equity capital is bound to the price of common stock (Bierman and Smidt, 1969). Difficulties of prediction in this area are well known.

Another overall source of uncertainty is the socioeconomic environment. Price controls, environmental restrictions, and tax changes are but a few of the important domestic factors. The level of imports is another factor which has a definite impact on the steel industry.

2. Endogenous Factors

Changes within a plant cause variations in profitability by modifying the production function—the relationship between output and inputs. The level of plant technology affects both the productivity of workers and the overall yield of the process. It is also intricately bound to the problem of scaling because different technologies are optimum at various levels of output. Fortunately, most plants are constructed with process equipment having well-established technology so that difficulties in estimation are mainly confined to the break-in period. After production begins. some cost reduction can be expected due to the learning curve as workers improve their efficiency with time. However, Abernathy and Wayne (1974) point out the pitfalls of relying on this effect.

E. Risk Analysis

Cash flow analyses have several shortcomings which limit their utility. The use of point estimates for each parameter dictates that only one of the many possible outcomes is determined. Sensitivity analyses serve to reveal several possible outcomes but no estimate of the likelihood of each is determined. Finally, the effects of change are often neglected, which can introduce serious errors in economic evaluations.

Many of the variables associated with a capital investment take on several values, some values more likely than others. Thus, each variable has an associated probability distribution which should be considered in a thorough analysis of a proposed project. Raiffa and Schlaifer (1968) describe some analytical techniques of incorporating probability distributions of variables into the determination of the probability of certain outcomes, but these methods become impractical as the number of variables increases. Monte Carlo methods such as risk analysis are well suited to solve such complex problems.

1. Outline of Techniques

Figure 2 shows the basic principles of risk analysis. The first step is to construct a model of the project under consideration; the model should contain all of the important variables of the process. At this point, the model should contain all of the important variables of the process. At this point, the model used for risk analysis is the same as would be needed for any

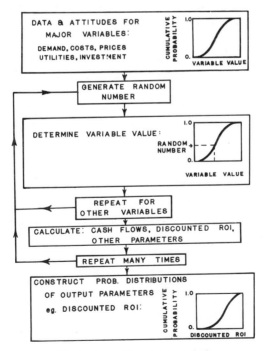

Fig. 2   An outline of risk analysis.

standard economic evaluation. The difference is that, in risk analysis, a probability distribution is substituted for each variable value while only the most likely value is used in conventional economic evaluations. In risk analysis, it is most convenient to use cumulative probability distribution functions, as defined by

$$F(X_k) = \text{Prob}(X \le X_k) = \sum_{i=1}^{k} \text{Prob}(X_i) \qquad (2)$$

where X is a discrete variable and the cumulative distribution functions are constrained by

$$0 \le F(X_k) \le 1 \qquad (3)$$

Once the probability distributions of each important variable have been obtained, the actual simulation begins. Figure 2 shows that the next step of risk analysis is to pick a specific value of each uncertain variable by random sampling. Most computers have available a program to generate uniform random numbers in the range from zero to one. Random number tables may be used in the absence of such a program. Equation (3) shows that the cumulative distribution function of each uncertain parameter also is constrained between zero and one. Thus, each random number corresponds to one and only one particular value of a variable, and a particular random number, say 0.527, corresponds to the particular value V_i of one variable of interest. This numerical value is saved for use in the model evaluation. In a similar manner, each other uncertain variable is given a randomly selected value.

After each variable has been assigned a numerical value, they are substituted into the economic model of the project to determine the cash flows for one year of the project life. Then new random numbers are generated and other numerical values assigned to the process variables. Again, cash flows for the next year are determined. After 16 repetitions of this procedure (corresponding to the 16 years of the project life) the complete cash flows of the project are determined. A discounted rate of return and present worth of the project may then be calculated by techniques which have been described in Section II.C.2. Thus, a complete cash flow analysis of one possible outcome has been completed.

It is obvious that other outcomes would have occurred had other random numbers been selected. Thus, the entire process must be repeated many times so that a statistically significant sample of possible outcomes is collected. It is this requirement that necessitates the use of a computer for risk analyses; one cash flow analysis calculation may require an hour of manual calculation.

In general, the exact number of repetitions needed depends on the form and number of the probability distribution of the variables and on the prescribed level of accuracy in estimating present values. Hahn (1972) gives a method to estimate the required number of repetitions in advance of the bulk of the calculations by using a procedure similar to that of estimating the binomial parameter from the results of an experiment. However, it is the experience of the authors and others that 500 to 1000 repetitions are sufficient to control the error of the estimate of the discounted rate of return to less than 0.2% and this requires less than three minutes of computation on a minicomputer.

2. Probability Distribution Construction

We now turn to the problem of constructing the cumulative probability distributions used in risk analysis. One method uses the first equality and requires a direct estimation of the probability that a variable will be less than or equal to a specific value. An alternative method used most often in practice is to make use of the second equality and to construct a frequency distribution of the variable values. The value of the cumulative distribution function can then be determined by summation.

It cannot be overemphasized that these probability distributions are subjective. The usefulness of risk analysis as a tool in decision making is completely dependent on the validity of the probability distributions. With this caveat in mind it may seem quite difficult to construct probability distributions, but this is not the case. Those who estimate the parameter values for use in typical economic evaluations often have great difficulty in deciding what variable value to use. This situation is most likely when wildly fluctuating variables such as scrap prices must be estimated. It is often easier to construct a histogram of possible values than to select one best value.

The method by which histograms are determined is completely dependent on its maker. Historical data may provide insight into some variables. Techniques such as multiple linear regression and econometrics (Wise, 1975) have been reported to be useful in practice. Implicit in these procedures is the assumption that past trends will continue into the future. Aside from the technical problems associated with estimating future variable values, Conrath (1973) offers some insight into the psychological aspects of probability distribution estimation.

One unique advantage of risk analysis is that correlations between variables or interactions are easily introduced into the model. For example, periods of rising steel prices are often associated with rising wages of steel workers. In risk analysis, this correlation is accounted for by using the same random number to assign values of wage rates and steel prices. Thus, steel prices and wages will tend to move in step.

3. Modeling Decision Strategies

In the previous sections it was implicitly assumed that all decisions related to the project under study would be made before the investment was undertaken. The probability distributions of the values of the variables were unconditional and not dependent on any event or decision. This may be too restrictive to adequately model certain situations. Suppose, for example, that scrap exports increase sharply so that scrap becomes unavailable or very costly. It is possible of course to account for this possibility by assigning a higher probability to higher scrap prices so that the distribution again becomes unconditional. In reality, if such an event were to occur, the plant managers could have several alternatives to paying higher prices. One possibility would be to substitute lower grades of scrap for higher grades to reduce cost, if the quality standard can tolerate the tramp elements. Another alternative would involve the use of direct reduced pellets. A temporary shutdown of the plant is also possible. The three alternatives to paying higher prices for scrap may produce drastic changes in the operation of the plant which cannot be adequately handled by numerical changes in the original model. For example, if low grade scrap were used, yield, internal quality, and surface quality would decrease, perhaps to a point where only hot-rolled material could be produced.

One solution to this problem is to have two or more models incorporated into the risk analysis representing each decision possibility. The first model would be used if scrap prices remained competitive. If a probability of soaring scrap prices were incorporated into the model, occasionally this event would occur, and the analysis would shift to an alternate model. It should be noted that many of the probability distributions would vary from model to model. In other words they would no longer be unconditional distributions but conditioned on a particular response to higher scrap prices.

When more than a few alternatives are available, this model becomes unwieldy. The technique of replacing the many decision branches by certainty equivalents, described by Schlaifer (1969), should be used to reduce the complexity of the problem. However, simple decision networks can be integrated into the risk analysis.

4. Sensitivity Analysis

The purpose of a sensitivity analysis is to show the results of a change in a project parameter. If the probability distribution of this parameter has been included in the risk analysis, the effect of the change, appropriately weighted by its likelihood, has already been included in the analysis. A sensitivity analysis may be useful to identify the relative significance of each uncertain variable. If there remains a feeling that the risk analysis does not span the expected range of outcomes, the original probability distribution must be modified.

5. Interpretation of Results

Interpretation of the results of a risk analysis is inherently bound to the value system of the decision maker. The form of the results consists of probability distributions of present worth, return on investment, and other parameters. Let us say that a risk analysis shows that the return on investment for a particular project varies from −20 to +50%, and we want to decide whether the investment should be undertaken. If we are very conservative we may decide that a negative rate of return would be disastrous and the project should be rejected however small the probability of occurrence. This is a maximin strategy because by not undertaking the investment, we have maximized our minimum gain. On the other hand, we may have a gambling instinct and decide to follow a maximax strategy to maximize our maximum gain. In this case the probability of a 50% return on investment leads us to accept the investment. Intermediate risk postures are also possible. One procedure is to calculate an expected rate of return by weighting each outcome by its likelihood. The decision is made to proceed with the project if the expected rate of return is greater than some specified minimum value. For small projects, most organizations are willing to decide on the basis of expected values; as the consequences of the various outcomes become greater, risk aversion increases. Most people would not be willing to risk all of their assets on a toss of a coin in a double or nothing gamble even though the expected value of the gamble is zero, the same value as nonparticipation. Conversely, if the wager were one penny, many would take the chance.

Modern utility theory, originated by von Neumann and Morgenstern (1947) and integrated into probability theory by Savage (1972), is often invoked to rationalize the behavior described above. We will not elaborate on the theory here other than to say that utility theory allows a decision maker's preferences to be mapped onto the scale of real numbers. After this has been done, all decisions should logically be made so as to maximize the expected utility of the outcomes. For most business decisions where outcomes are expressed in dollars, it often suffices to merely assign weights to the various money outcomes of a decision and use that as a utility. Thus, a positive one million dollar outcome in present value could be assigned a value of 1.0 while a two million dollar outcome might become only 1.5. A negative one million dollar outcome could become −1.0. Our decision would then be based on maximizing the expected value of these utilities.

III.   THE PRODUCTION OF STEEL STRIP FROM POWDER

A.   Process Outline

A P/M steel plant is to be designed to produce sheet from scrap consistent with the metallurgical constraints of the process and utilizing present operating practices where possible. To minimize interactions of the proposed plant with scrap and finished steel prices, the scale of operation is taken as 175,000 tons/yr (157,000,000 kg/yr). The location is hypothesized to be on a green-field site in the Chicago–Detroit area.

A material flow diagram of the process is shown in Fig. 1. The production route is designed to produce hot-rolled band 48 in. (1.2 m) in width at an average thickness of 0.1 in. (2.5 mm). This gage and width is sufficient to supply much of the current requirements for steel sheet (U.S. Steel, 1964; Kuhl, 1972a,b,c,d). Hot-rolled steel can be produced in gages from 0.135 to 0.050 in. (3.44–1.27 mm) on the same equipment. Cold-rolled sheet is fabricated from the hot band in gages from 0.105 to 0.030 in. (2.67–0.76 mm). The composition would generally be that of conventional low carbon steel and be controlled by careful selection of scrap and conventional melting practices.

The plant capacity is 220,000 tons/yr (198,000,000 kg/yr), but it is estimated that it will operate at a utilization of 80% to give an average production rate of 175,000 tons/yr (157,000,000 kg/yr). Twenty 8-hr shifts/week and 48 weeks/yr of operation are assumed. The idle time would be used for maintenance and major repairs.

The steel sheet is produced in three forms: coil, sheet, and strip. Selling prices, scrap prices, and general plant parameters are shown in Table I. The following sections describe each of the process steps in detail.

B.   Technical Details

1.   Powder Production

The production of powder by atomization is the first step in the P/M steel plant. Gregory and Bridgwater (1968) reviewed this and other existing and proposed processes for powder production. Although they indicated the

TABLE I

General Plant Parameters

Parameter	Value	Reference
Process building	4,050,000 ft^3 at \$1.80/ft^3	Jelen (1970)
Warehouse building	900,000 ft^3 at \$1.31/ft^3	Jelen (1970)
Repair and lab building	300,000 ft^3 at \$5.51/ft^3	Jelen (1970)
Offices	10,000 ft^2 at \$37.80/ft^2	Jelen (1970)
Land	210 acres at \$3000/acre	
Fencing	12,000 ft^2 at \$13.50/ft	Jelen (1970)
Roadways and parking	72,000 ft^2 at \$1.40/ft^2	Jelen (1970)
Railroad siding	1,000 ft^2 at \$37.80/ft	Jelen (1970)
Wage rate	\$10.59/hr	AISI (1976)
Salary rate	\$12.05/hr	AISI (1976)
Electricity	\$0.013/kWhr	
Natural gas	\$1.50/1000 ft^3	
Water	\$0.20/1000 gal	
No. 1 scrap price	\$0.0286/lb	*Iron Age* (1976)
No. 2 scrap price	\$0.0191/lb	*Iron Age* (1976)
Hot-rolled steel price	\$0.1170/lb	*Iron Age* (1976)
Cold-rolled steel price	\$0.1390/lb	*Iron Age* (1976)

possibility of producing low-cost powder from low-grade scrap or ore by hydrometallurgical methods, these have not been demonstrated on a commercial scale and so were not used in the design of the P/M steel plant.

Scrap and other materials shown in Fig. 1 are melted in two 55-ton (50,000 kg) electric furnaces with 20,000 kVA transformers having a tap-to-tap time of 3.1 hr (*33 Magazine*, 1974). High-grade or No. 1 scrap is the principal feedstock for melting although lower grade No. 2 bundles would constitute an estimated 10% of the scrap charge. High levels of copper and tin in No. 2 bundles increase the risk of hot shortness and surface defects.

Other raw materials charged into the electric furnaces are coke, lime, and fluorspar. Kuhl (1972a) estimates that they are required in the amounts of 1, 4, and 0.1% of the metallic charge, respectively. Other parameters of melting are shown in Table II.

After the charge is melted and refined, it is tapped into the tundish of an atomizer. The atomization process is still somewhat of an art with the technology established from operational and sometimes proprietary data. However, some information has been published as patents or elsewhere. Gummeson (1971) has reviewed atomization and relates the process variables to powder characteristics. In-depth study of these relationships is given in Chapter 1.

Gas atomization with noninert gases tends to produce powders which

TABLE II

Details of Melting

Parameter	Value	Reference
Capacity (tons/year)	270,000	
Yield (%)	92	Kuhl (1972a)
Scrap (%)	1	Kuhl (1972a)
Capital cost, inclusive of auxiliaries	$8,750,000	Kuhl (1972a); Cartwright (1972)
Electricity consumption (kWhr/ton)	500	*33 Magazine* (1974)
Maintenance (% of capital cost)	10	Kuhl (1972a,c)
Electrode cost ($/ton)	$7.30	*33 Magazine* (1976)
Hourly workers/shift	12	Kuhl (1972d)
Salaried workers/shift	5	Kuhl (1972d)

are irregular spheres. Water atomization has been favored by Gregory and Bridgwater (1968) and Gummeson (1971) to produce carbon steel powder. The resulting irregularly shaped particles impart a high green or unsintered strength to powder compacts. Surface oxides formed during water atomization must be removed by annealing in a reducing atmosphere. Annealing also softens the powder for better compaction.

The production rate of the P/M steel plant requires two atomizers, each of 125,000 tons/yr (112,000,000 kg/yr) capacity. They are to be designed to use either gas or water as an atomization agent. A change of nozzles would be necessary to effect the change from gas to water. It is expected that water atomization will be used for low carbon steel, while gas atomization may be needed to produce other alloys (see Chapter 1).

The maximum powder particle size acceptable for powder rolling is about 20 to 32 mesh (0.84–0.51 mm) although thin strip requires finer powders. The large maximum particle size of the powder permits a high yield to be attained during atomization. A distribution of powder sizes is produced by atomization. Such distributions of powder sizes have been successfully powder rolled provided that the amount of fines is not large. See Sturgeon 1968, 1972), Gummeson (1971), Shakespeare (1968), and Storchheim (1956) for details of the relationship between atomization and powder rolling. Table III gives details of atomization and powder annealing.

After atomization, the powder is dried and sieved to remove oversize particles. Then the powder is annealed under an endothermic atmosphere in a continuous-belt electrically-fired furnace for 1 min at 1340°F (725°C).

2. Powder Rolling

Powder rolling is the technique most often proposed for strip production from powder because of its continuous nature and high potential output.

TABLE III

Details of Atomization and Powder Annealing

	Atomization	Powder annealing
Capacity (tons/year)	225,000	240,000
Yield (−32 mesh) (%)	95	100
Capital cost	$1,550,000[a]	$2,050,000[b]
Electricity (kWhr/ton)	260[a]	100
Hourly workers/shift	14[a]	3
Salaried workers/shift	2	—
Maintenance (% of capital cost)	10	10

[a]Shakespeare (1968).
[b]Ascough (1968).

Tracey (1969) has published a review of the process with an extensive bibliography. Powder rolling consists of feeding powder into the roll nip of a light rolling mill and allowing friction of the rolls to drag the powder into a progressively narrow gap. Initial densification occurs by particle re-arrangement. As rolling proceeds, the density of the strip increases due to plastic deformation and cold welding of the powder particles. The powder rolled strip is immediately fed into a sintering furnace on a roller conveyor to increase its strength. Successful high-speed powder rolling at 40–100 ft/min (10–30 m/min) depends on control of powder flow, strip speed, and roll gap (Tundermann and Singer, 1968, 1969). An edge control system involving side rolls or flanges is necessary to maintain strip density and strength (Tracey, 1969).

The thickest hot-rolled strip to be produced for sale is 0.135 in (3.4 mm) in thickness, so that allowing for a minimum 50% reduction during hot rolling, the powder rolled strip would need to be 0.27 in (6.9 mm) thick. A light rolling mill with 40 in (1 m) diameter rolls 49 in (1.25 m) wide could meet these specifications. Table IV gives details of powder rolling and sintering.

3. Sintering

The strip produced by powder rolling is weak and difficult to coil. In the P/M steel plant it would be fed directly into a horizontal sintering furnace. Because the strip has a velocity of tens of feet per minute, it is necessary to use a short sintering time to minimize the size of the sintering furnace. Thus, the highest possible temperature must be used. Tracey (1969) and Sturgeon (1968) have shown that 5 min at 2190°F (1200°C) are required. Even higher temperatures produce only small improvements in strip properties but create severe furnace design problems. If we assume a strip

TABLE IV

Details of Powder Rolling and Sintering

	Powder rolling	Sintering
Capacity (tons/year)	500,000	250,000
Yield (%)	98	100
Capital cost	$1,350,000	$4,000,000
Electricity (kWhr/ton)	8	306
Hourly workers/shift	3	2
Salaried workers/shift	1	—
Maintenance and rolls (% of capital cost)	35	10

speed of 40 ft/min (12 m/min), the size of the sintering furnace becomes 270 ft (82 m) allowing for heat-up time. A protective atmosphere is required both to prevent oxidation and control carbon content.

4. Hot Rolling

Although cold reduction after sintering is possible, hot rolling is more efficient because it eliminates several intermediate annealing steps (Crooks, 1962). Again, a protective atmosphere is required to minimize oxidation. Picking of the strip becomes unnecessary.

The first hot-rolling pass must be greater than 20% reduction to minimize transverse cracking of the strip and to increase the strip density from 80 to 90% of theoretical. Two additional hot passes raise the density to 100% of theoretical and yield a strip virtually identical in mechanical properties to conventional hot band (Crooks, 1962).

A 4-high reversing mill with coilers at each end covered by furnaces to maintain the temperature is used for hot rolling. Rolling sintered strip without coiling is impractical because the slow strip speed causes overheating of the rolls. Table V gives details of hot rolling.

5. Finishing

Because the hot-rolled P/M strip is assumed to be similar to conventional hot band, the process of final cold reduction and annealing is the same as is used for conventional material.

Cold reduction could be accomplished on a 4-high reversing mill which would double as a temper mill (Shakespeare, 1968). Box annealing would require three conventional furnaces with eleven annealing bases (Ascough, 1960).

TABLE V

Details of Hot Rolling, Cold Rolling, and Annealing

	Hot rolling	Cold rolling	Annealing
Capacity (tons/year)	500,000	230,000	100,000
Yield (%)	93	98	100
Capital cost	$9,500,000	$5,400,000	$1,300,000
Electricity (kWhr/ton)	16	98	240
Hourly workers/shift	5	4	3
Salaried workers/shift	1	1	—
Maintenance (% of capital cost)	15	10	10
Rolls (% of capital cost)	40	13	10

C. Basic Financial Model

Even the most rudimentary economic analysis of a P/M steel plant contains an estimate of the capital needed and the resultant operating costs and revenue. Profit is obtained as the difference between revenue and cost and is divided by the initial investment to obtain a rate of return on investment. In this section, elements of the estimation of capital and operating costs for the P/M steel plant will be described. These principles were used to derive the costs in Tables II–V.

A summary of capital cost estimation is shown in Table VI. Usually the cost of the major equipment is determined first and installation, piping, and electrical auxilliaries are estimated as a percentage of that value. Jelen (1970)

TABLE VI

Elements of Capital Cost Estimation

Total investment
  Working capital
  Permanent investment
    Land
    Depreciable investment
      Site improvements
      Buildings
      Installed equipment cost
        Major equipment items
        Installation
        Electrical auxiliaries
        Piping
        Environmental control
        Materials handling

suggests 40, 2, and 33%, respectively. The best sources of data for the cost of major equipment are quotations from suppliers, but less reliable and out-of-date published data must often be used. The Chemical Engineering Plant Cost Index (CEPCI) shown in Fig. 3 is used to update published plant costs to the present by

$$\text{current cost} = \text{published cost} \times \frac{\text{CEPCI (current year)}}{\text{CEPCI (year of publication)}} \qquad (4)$$

Many capital and operating costs are not indexed so that a very broad measure of cost, the GNP price deflator shown in Fig. 3, is used instead. Costs of similar type, but different capacity equipment, are adjusted by

$$C_2 = C_1 (Q_2/Q_1)^x \qquad (5)$$

where C_1 is the known cost of capacity Q_1, C_2 the desired cost of capacity Q_2, and X the cost capacity factor. This relationship is often referred to as the six-tenths rule because X is usually 0.6. Equation (5) has a profound influence on the scaling of steel plants, a subject discussed thoroughly by Leckie and Morris (1968).

Working capital is needed to maintain necessary inventory, raw materials, cash, and accounts receivable levels. Working capital is approximately 20% of annual sales if one month of inventory, cash, etc., are kept on hand. At the end of the project life all working capital is recovered by the running down of cash and stock of materials. Table I lists typical values for other plant parameters.

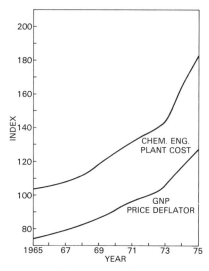

Fig. 3   Cost indices.

TABLE VII

Elements of Operating Cost and Revenue

Sales revenue
  Cost of sales
    Mill cost
      Processing costs
      Raw materials
      Indirect labor
      Depreciation
    Overhead cost
      Selling Costs
      Insurance
      Administration
      Property taxes
  Gross profit
    Federal income tax
    Net profit

Operating cost and revenue estimation, another basic part of an economic analysis, is summarized in Table VII. Sales revenue is the product of demand (nominally 175,000 tons/yr (157,000,000 kg/yr)) and published prices for hot- and cold-rolled steel (*Iron Age*, 1976). Figure 4 shows recent trends in prices and wages. Processing costs are the sum of labor, utilities, and maintenance costs. Details are given in Section III.B, where specific operations are described. Raw material costs are based on published prices for scrap

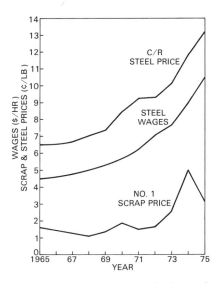

Fig. 4 Recent trends in scrap and steel prices and wages.

and other materials (*Iron Age*, 1976; Kuhl, 1972a,b,c,d). Indirect labor, estimated at 10% of direct labor, includes quality control, testing, and labor for environmental controls. Depreciation, an accounting expense requiring no cash outlay, is normally shown on a straight-line basis over a standard 14-yr period, although more advantageous accelerated methods are often used for tax purposes (Bierman and Smidt, 1969). Overheads are usually calculated on the basis of a percentage of sales or investment. Selling costs, which include freight equalization allowances, are estimated at 5% of sales revenue, while administration is assumed to be 1% of revenue. Insurance costs are based on 2% of the depreciable investment, and property taxes are 2% of the total permanent investment. Tables VIII and IX summarize the capital and operating details of the P/M steel plant.

IV. RISK ANALYSIS OF A P/M STEEL MINIPLANT

A. Construction of Probability Distributions

1. Capital Costs

One major source of risk in undertaking a project is the uncertainty in the estimate of the capital investment. This type of error usually results in under-estimation. It generally occurs because certain items of equipment are over-looked in the original planning for the project. If we assume that the P/M steel

TABLE VIII

P/M Steel Plant Capital Requirements

Process equipment	$34,650,000
Installation	13,860,000
Electrical auxiliaries	8,620,000
Process piping	693,000
Environmental equipment	1,530,000
Materials handling	1,350,000
Installed equipment cost	$60,703,000
Land improvements	300,000
Building costs	$11,420,000
Depreciable investment cost	$72,423,000
Land	630,000
Permanent investment	$73,053,000
Working capital	9,026,000
Total investment	$82,079,000

TABLE IX

P/M Steel Plant Operating Summary

Hot-rolled sales	$18,720,000
Cold-rolled sales	26,410,000
Total sales	$45,130,000
Operating process costs	19,340,345
Raw materials	10,942,115
Indirect labor	395,856
Depreciation	5,173,071
Mill cost	$35,851,387
Selling cost	2,256,500
Insurance	1,448,460
Administration	451,300
Property tax	1,461,060
Overhead cost	$ 5,617,320
Cost of sales	$41,468,707
Gross profit	3,661,293
Federal income tax	1,757,420
Net profit	1,903,873
Return on investment	2.6%

plant proposal has been carefully prepared, the estimated capital cost should be biased only slightly toward the lower values. Therefore, it is estimated that Table X adequately represents our capital cost inaccuracy factor.

2. Start-Up Costs

Start-up costs were estimated to be 10% of the installed equipment cost in the cash flow analysis. Jelen (1970) states that the 10% figure is seldom exceeded. While 10% is the most likely value for start-up, a range of values is possible. The novelty of the process should cause more difficult start-up than usual, but this factor may be offset by the relative simplicity of the process. Based on these considerations Table XI was constructed to represent start-up costs.

TABLE X

Probability Distribution of Capital Cost Inaccuracy

Actual capital cost ÷ estimated capital cost	0.9	1.0	1.25
Cumulative probability (%):	25	60	100

3. Selling Prices for Steel Sheet

It is assumed that P/M steel sheet will sell for the same price as conventional materials, or the published price in *Iron Age* magazine. It is very difficult to forecast the price trend for the next decade. Increasing capital requirements in the industry should exert a strong upward pressure on prices, particularly if imports of steel do not increase. On the other hand, competition between materials in many uses is very keen. Without any firm conviction, Table XII was constructed as an estimate of steel price trends. This table duplicates the price changes observed during the period 1965–1975.

4. Demand

The P/M steel plant would probably be built in a market with a growing demand for steel; a 3% annual market growth rate was assumed. However, competitive factors and the state of the economy could cause changes in the demand. The P/M steel plant should have a higher operating rate than the overall industry because the P/M process has a much shorter time constant. Another marketing advantage is that it can supply relatively small customers directly instead of through a service center. This would offer customers a price advantage. Table XIII is the probability distribution of demand as a percentage of the average demand given by the 3% trend line.

TABLE XI

Probability Distribution of Start-Up Costs

Start-up costs (% of installed equipment cost):	5	6	8	9	10	12
Cumulative probability (%):	10	20	40	60	90	100

TABLE XII

Annual Changes in the Price of Steel Sheet

Annual price change (%):	2	4	5	6	7	18	26
Cumulative probability (%):	20	40	50	60	80	90	100

TABLE XIII

Demand for Steel as a Percentage of Average Demand

Demand as a percentage of average demand:	80	85	90	100	105
Cumulative probability (%):	50	60	70	90	100

5. Scrap Prices

Very few prices are more volatile than the price of scrap as shown in Fig. 4. Even long-term trends are difficult to estimate because scrap prices are affected by many long- and short-term conditions. Obviously, local supply and demand conditions determine the short-range price. Small changes in demand from higher steel production or greater exports can have important effects. In the long term, the balance between the number of electric furnaces and basic oxygen furnaces (BOF's) will be a key factor. Other trends may be caused by the growth of direct reduction or scrap pre-heating for the BOF. Environmental attitudes may also have an important role to play. Lacking a model such as Wise (1975) has presented, it has been assumed that future gyrations of scrap prices would follow the pattern of the period 1965–1975. Table XIV shows the distribution of scrap prices as a percentage of its price the previous year. A constraint was imposed that the scrap price could not exceed 60% of the selling price for steel sheet.

6. Labor Costs

The probability distribution of annual wage increases was constructed on the basis of actual labor agreements from 1965 to 1975. Table XV shows these probabilities. This type of model is unrealistic because it fails to adjust for productivity changes and demand effects. The effect of demand for steel on labor costs was accounted for by assuming a Cobb–Douglas production function for steel (Intriligator, 1971). This function assumes a power relationship between output and each input. The exponent was chosen to be 0.25 for labor. Productivity changes were accommodated by assuming that the plant capacity followed a 90% learning curve (Abernathy and Wayne,

TABLE XIV

Annual Changes in Scrap Prices

Scrap price as a % of its price the previous year:	63	80	86	88	92	108	123	139	159	195
Cumulative probability (%):	10	20	30	40	50	60	70	80	90	100

TABLE XV

Annual Wage Rate Increases

Annual wage increases (%):	3	6	7	8	10	13	16	18
Cumulative probability (%):	20	40	50	60	70	80	90	100

1974). In other words, only 90% as much labor per unit of output is required as the cumulative output doubles.

7. Electric Power Costs

Historical electric power costs are probably irrelevant for predicting future costs. For many years electric power rates actually declined in current as well as real dollars. At present, higher rates seem assured. Table XVI shows the probability distribution of the annual increase in electric power rates. Again a Cobb–Douglas production function was assumed with a coefficient of 0.9.

8. Maintenance Costs

Maintenance costs were estimated for each item of equipment in Section III. Recognizing that these were just estimates, the probability distribution in Table XVII reflects our view that much uncertainty exists.

9. Taxes

The federal income tax rate of 48% was used in the cash flow analysis. For the risk analysis, it was assumed that taxes would tend to become lower, although not markedly. Table XVIII shows the assumed probability distribution of different tax rates.

10. Other Factors

All operating costs such as rolls, maintenance, and environmental which are not indexed were assumed to increase at the same rate as the GNP deflator, the broadest measure of prices in the economy. The trend of the

TABLE XVI

Annual Increase in Electric Power Rates

Annual increase (%):	0	2	4	5	6
Cumulative probability (%):	10	30	70	90	100

TABLE XVII

Maintenance Probability Distribution

Maintenance cost as a percentage of that assumed in Section II:	60	70	80	90	100	110	130
Cumulative probability (%):		10	20	30	50	80	90	100

TABLE XVIII

Federal Income Taxes

Corporate income tax rate (%):	43.2	45.6	48.0	50.4
Cumulative probability (%):	20	50	90	100

GNP deflator shown in Fig. 3 was assumed to continue into the future. The probability distribution is given in Table XIX.

B. Alternate Formulations

1. Role of Historical Data

One of the flexibilities of risk analysis is that the role of historical data can be important or negligible according to the beliefs of the analyst. In our discussion of the probability distributions used in the P/M plant it was noted that some were based on historical data and others were not. There is just no objective method for making the decision.

2. Effects of Interaction

In the P/M steel plant risk analysis, the GNP price deflator, steel prices and wages, scrap prices, and electric power rates were correlated by using the same random number for each. An alternative method of introducing an interaction is to explicitly relate two variables in one of the equations defining the plant model. After this has been done, only a single probability distribution of either of the variables is needed. The value of the other is determined from the equation.

One type of interaction most difficult to model is that involving the potential actions of a competitor, particularly if he is able to react to any of your actions. It is possible to describe a simple one-competitor game, but more than one competitor creates an intractable problem. Competitive effects have not been considered explicitly in the P/M steel plant model.

TABLE XIX

Rate of Increase of the GNP Price Deflator

Rate of increase (%):	3	4	5	6	9	11
Cumulative probability (%):	20	40	70	80	90	100

C. Results

1. Discounted Cash Flow Analysis

Figure 5 shows the results of the cash flow analysis of the P/M steel plant in the form of a present value prediction. The discounted rate of return is less than any cost of capital that could be reasonably assumed. In other words, the project should not be accepted. Note that the 5.9% is higher than the 2.6% shown in Table IX due to the positive effects of accelerated depreciation and the investment tax credit.

2. Risk Analysis

Figures 6–8 show the cumulative probability distributions of annual raw materials costs, labor costs, and sales revenue over the 16 year life of the P/M steel plant project. Each curve on the figures is drawn at a certain percentile of the distribution. The 50% curve represents median values.

All of the curves have a positive slope with increasing time and also an increasing variance with time. The slopes are easily interpreted in terms of inflation and somewhat by the learning curve effect on output with time. It also seems quite reasonable that uncertainty should increase in the more distant future. Labor costs appear to be the most predictable, and scrap prices the least predictable.

Figures 9 and 10 show the frequency distribution and cumulative probability distribution of the return on investment. Note that the distributions have a large range in the lower rates of return but a sharp cutoff at the upper end. In contrast to the results of Malloy (1971), the P/M steel plant risk analysis projects a higher rate of return than a cashflow analysis. The median return is about 8%, and there is a greater than two-thirds chance that the predicted cash flow analysis rate of return will be achieved.

Figure 11 shows the cumulative probability distribution of present worth above 7%. The probability of a positive present worth appears to be about 60%, which is in rough agreement with the rate of return results.

D. Risk Analysis and the Future of P/M Steel Strip Production

A few definite conclusions about the future of P/M steel plants can be deduced from this study. The profitability of a P/M steel plant appears to be barely adequate but with a substantial risk of a poor rate of return. Few organizations will be encouraged by this study to build a P/M steel plant.

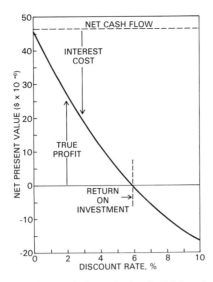

Fig. 5   Cash flow analysis results for the P/M steel plant.

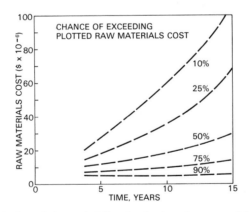

Fig. 6   Cumulative probability distribution of raw material cost.

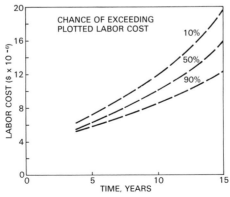

Fig. 7   Cumulative probability distribution of labor costs.

201

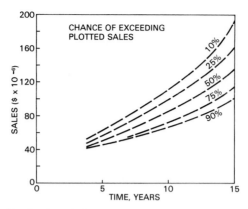

Fig. 8 Cumulative probability distribution of sales.

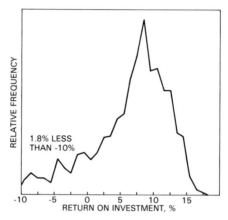

Fig. 9 Frequency distribution of return on investment.

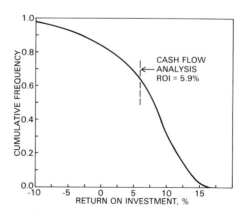

Fig. 10 Cumulative probability distribution of return of investment.

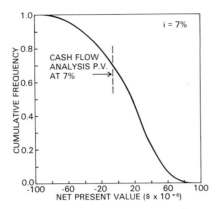

Fig. 11 Cumulative probability distribution of present value.

It seems that the profitability of the P/M plant is only slightly greater than for a conventional one but involves a higher risk.

Lower cost powder produced from ore or low grade scrap by hydrometallurgical methods appears to hold the key for the future of P/M steel plants.

REFERENCES

Abernathy, W. J., and Wayne, K. (1974). *Harvard Bus. Rev.* **52**, 109–119.
Ackoff, R. L., and Sasieni, M. W. (1968). "Fundamentals of Operation Research," pp. 32–43. Wiley, New York.
AISI (1976). "AISI Annual Statistical Report." American Iron and Steel Institute, Washington, D.C.
Ascough, H. (1960). "Production of Wide Steel Strip." Iron and Steel Institute, London.
Bierman, H., and Smidt, S. (1969). "The Capital Budgeting Decision." Macmillan, New York.
Cartwright, W. F. (1972). *J. Iron Steel Inst. (London)* **210**, 221–228.
Conrath, D. W. (1973). *Management Sci.* **19**, 873–883.
Crooks, S. R. (1962). *Iron Steel Eng.* **39**, 72–76.
Gregory, S. A., and Bridgwater (1968). *Powder Met.* **11**, 233–260.
Gummeson, P. V. (1971). "Modern Atomizing Techniques." Worcester Polytechnic Institute Worcester, Massachusetts.
Hahn, G. J. (1972). *IEEE Trans. Syst. Man Cybern.* **SMC-2**, 678–680.
Hertz, D. B. (1964). *Harvard Bus. Rev.* **42**, 95–106.
Intriligator, M. (1971). "Methematical Optimization and Economics Theory." Prentice-Hall, Englewood Cliffs, New Jersey.
Iron Age (1976). *Iron Age* **217**, 107, 109.
Jelen, F. C. (1970). "Cost and Optimization Engineering." McGraw-Hill, New York.
Klein, L. R. (1975). *Proc. IFAC 6th World Congress*, pp. 5:1–5:8.
Kuhl, R. J. (1972a). *J. Metals* **24**, 41–48.
Kuhl, R. J. (1972b). *J. Metals* **24**, 40–44.
Kuhl, R. J. (1972c). *J. Metals* **24**, 39–44.

Kuhl, R. J. (1972d). *J. Metals* **24**, 28–30.

Leckie, A. H., and Morris, A. J. (1968). *J. Iron Steel Inst.* (*London*) **206**, 442–452.

Luce, R. D., and Raiffa, H. (1957). "Games and Decisions: Introduction and Critical Survey." Wiley, New York.

Malloy, J. B; (1971). *Chem. Ind.* (*London*) **49**, 1242–1250.

Mitroff, I., and Betz, F. (1972). *Management Sci.* **19**, 11–24.

Mo, W. Y., and Wang, K. L. (1970). "A Quantitative Economic Analysis and Long-Run Projections of the Demand for Steel Mill Products." Bureau of Mines, Washington, D. C.

Raiffa, H., and Schlaifer, R. (1968). "Applied Statistical Decision Theory." Massachusetts Institute of Technology Press, Cambridge, Massachusetts.

Regan, W. J., James, R. W., and McLeer, T. J. (1972. "Identification of Opportunities for Increased Recycling of Ferrous Solid Waste." PB-213577, Institute of Scrap Iron and Steel.

Savage, L. J. (1972). "The Foundations of Statistics." Dover, New York.

Schlaifer, R. (1969). "Analysis of Decisions Under Uncertainty." McGraw-Hill, New York.

Schuhmann, R. (1952). "Metallurgical Engineering." Addison-Wesley, Reading, Massachusetts.

Shakespeare, C. R. (1968). *Powder Met.* **11**, 379–399.

Storchheim, S. (1956). *Metal Progr.* **70**, 120–126.

Sturgeon, G. M. (1968). *Powder Met.* **11**, 314–329.

Sturgeon, G. M. (1972). *Sheet Metal Ind.* **49**, 59–25.

33 Magazine (1974). *33 Mag.* **12**, 33–43.

33 Magazine (1976). *33 Mag.* **14**, 42.

Tracey, V. A. (1969). *Powder Met.* **12**, 598–612.

Tundermann, J. H., and Singer, A. R. E. (1968). *Powder Met.* **11**, 261–294.

Tundermann, J. H., and Singer, A. R. E. (1969). *Powder Met.* **12**, 219–242.

U.S. Steel Corporation (1964). "The Making, Shaping, and Treating of Steel," 8th ed.

Varga, J., and Lownie, H. W., (1969). "A Systems Analysis of the Integrated Iron and Steel Industry." PB184577, Battelle Memorial Institute, IV-17-18.

von Neumann, J., and Morgenstern, O. (1947). "Theory of Games and Economic Behavior." Princeton University Press, Princeton, New Jersey.

Williams, E. T., and Johnson, R. C. (1958). "Stoichiometry for Chemical Engineers." McGraw-Hill, New York.

Wise, K. T. (1975)., *Iron Steelmaker* **2**, 23–32.

INDEX

A
B
C 8
D 9
E 0
F 1
G 2
H 3
I 4
J 5